Frances Hamerstrom

F R A N C E S   H A M E R S T R O M

BIRDING WITH A PURPOSE

Of Raptors, Gabboons, and Other Creatures

Drawings by Jack Oar

Jack Oar

The Iowa State University Press / Ames, Iowa

BOOKS BY FRANCES HAMERSTROM

The Adventures of the Stone Man

Walk When the Moon Is Full

An Eagle to the Sky

Strictly for the Chickens

Birds of Prey of Wisconsin

Birding with a Purpose

© 1984 Frances Hamerstrom. All rights reserved
Composed and printed by The Iowa State University Press

First edition 1984

Library of Congress Cataloging in Publication Data

Hamerstrom, Frances, 1907–
    Birding with a purpose.

    1. Hamerstrom, Frances, 1907–   .   2.   Bird-banding—United States. 3.
Ornithologists—United States—Biography.   I.   Title.
QL31.H25A33   1984   598′.092′4 [B]   83–12684
ISBN 0–8138–0228–8

CONTENTS

FOREWORD

This book consists of the behind-the-scene recollections of a raptor-trapper named Fran. She does not pronounce it Fran as in brand — the way my South Bronx background would have it, but Fron as in frond — the way they do in Back Bay, her point of origin in this life.

YOU'VE heard of wolf trappers, fox trappers, muskrat trappers, and the like. Raptor trappers are different. Officially, they want to band birds to learn about their weight and moult, their later movements, their longevity, and all that. Underneath, they are unabashed admirers of the wildness, magnificent strength, and awesome flight of creatures at the top of the animal pyramid. I wouldn't call them childlike; but they do have a youthful zest, and they will endure any hardship and go to any length to catch their birds. As this book will reveal. Fran is one of these youthful persons. In her own light-hearted way, she portrays here the fascinating development of raptor trapping techniques in the past few decades, a development in which she played a leading role.

This is a far from stuffy book. Sometimes the reader is a bit shocked, but more often explodes with laughter. Painlessly, he has learned new facts about birds, traps, and history.

I think it will help you, the reader, to know a little more about Fran. She met Frederick at a Dartmouth Zeta Psi Fraternity House Party. Not the place to start up a lasting friendship. But they remain completely in love after five decades. That's almost awesome. No one that I know has ever seen Frederick mad, except the time a young editor of the *Auk* tried to improve the Hamerstrom prose in an ornithological manuscript Frederick had submitted for publication. That was about 30 years ago, and the editor's face is still red.

The Hamerstroms, as very young neophytes, got into game conservation via a school of "vermin"-controllers in New Jersey in 1931; but they soon happily escaped the East to become students of Paul L. Errington at Iowa State University and then of Aldo Leopold at the University of Wisconsin, and they have been staunch ecologists ever since. Their main research for long was on prairie chickens, mostly for the Wisconsin Conservation Department; but Fran always had a side interest in raptors, and this became a full-time one in later years. The Hamerstroms' magnificent dedication to the declining prairie chickens of Wisconsin led Frederick and Fran to astute analyses of the ecological problems facing this species in that region and then to perceptive management plans that arrested the decline there and importantly increased prairie chicken numbers. It was indeed a memorable achievement in wildlife management and conservation.

Fran may be the last living relic of the era of ornithologists who started out with a shotgun instead of binoculars. That was Audubon's era too. She has authored or coauthored a bulletin on raptors, two books dealing with raptors, and sixty-one articles and technical papers on raptors. She has thus been in a marvelous position to trace the development of modern trapping techniques, the history of raptor trapping, and the recent changing of public attitudes toward our birds of prey. That is what this book is all about.

Birding with a Purpose has involved people who in other books and articles fascinating to the scientific community concentrated on song sparrows, Kirtland's warblers, bald eagles, red-cockaded woodpeckers, and the bobwhite quail. Dr. Hamerstrom's birding, the direct handling of a wide variety of raptors for scientific study, is equally purposeful—but is laced with rowdy zest.

Joseph J. Hickey

Department of Wildlife Ecology
The University of Wisconsin-Madison

viii

THIS BOOK is dedicated to those rugged, persistent individuals—both amateur and professional—who, without pay, band birds to enhance our understanding and further conservation, and to Frederick Hamerstrom who has shared most of my adventures.

.I never thought I'd dedicate a book to a government agency, but it is also dedicated to the Bird Banding Laboratory of the United States Fish and Wildlife Service for its splendid work.

An old-fashioned background

Bird watching—for me—tends to be with a purpose. As a child I used to borrow my mother's pearl-handled opera glasses, and climb high into the tops of the oaks and the hickories to see what the warblers were doing. My favorite oak held a small pool in a hollow crotch, high, high above the ground after each rain. I found myself a perch still higher so I could look down at my pool and watch the birds come to drink. They sometimes fought and—as I watched them day after day—a pattern developed. Some birds were king and could have the pool anytime they wanted. Others fought for their turn at the pool, and still others only slipped to the pool when no other birds were around. I wondered why. Were certain individuals more domineering?

A screech owl roosting in a rhododendron was discovered by a blue jay, and more and more jays came screeching and diving to mob it. As far as I knew nobody had ever watched anything like this before. This was the day I started to write my first bird book. Chapter I: How the jays hated the screech owl. Other chapters followed. Knowing nothing about plagiarism, if I found a good chapter for my book in a magazine, I cut it out and pasted it in. Sometimes I pasted my own handwritten account right over someone's published story, convinced that mine was right and his account was wrong. In retrospect I now see I was beginning to make progress as a scientist, for I no longer believed everything that was printed . . . a valuable lesson that some grown-ups have not learned yet!

My early problems with bird identification were stupendous. The only book I owned that could help me at all was *The Look About You Nature Book*. It was beautifully illustrated with vivid colored pictures; but this charming volume was British, and no one pointed out to me that the birds of Great Britain are not the same species we have in the United States!

I kept my birding increasingly secret. I hated being laughed at whenever I called crows "rooks" or referred to chicadees as "tits."

3

Besides, before long, I managed to get a BB gun and started collecting birds. This came about after my governess took me to a museum on one of our daily walks. She let me take a dead bird (found by the roadside) with me for identification. The men in the museum were delighted. They praised me, and they showed us the workroom where the stuffing of birds in lifelike positions was done. A small hunchback was putting the final touches on an exquisite thrush standing on a twig and singing with its mouth open. I leaned so close to see better that one of my long braids fell over his shoulder. He grunted and pushed me aside. I apologized and my governess took me away. It would have surprised me to even suspect that the hunchback and I were about to become fast friends. What I did know was that I needed dead birds for taxidermy, and for a way to get back to this wonderful Ward's Natural History Museum in Hyde Park—just a mile from our house. It never occurred to me that they might let me in the door, even if I had the temerity to arrive without at least one dead bird.

Thereafter I visited the museum alone, escaping from adult supervision on my bicycle. Something told me that my family would not understand if they found out (1) that I had a gun, and (2) that I had a hunchbacked friend who traded me glass eyes to put in the birds I mounted for *my* private collection in exchange for specimens of rarer birds that the museum wanted.

I learned to identify the birds of Massachusetts by wandering around in that museum studying labeled specimens, and I felt it my duty to supply them with as many specimens of the birds on their "want list" as I possibly could.

My family graciously gave me a vacant maid's room for my hobbies. It contained my insect collection, my mammal collection, my bird collection, my egg collection, arsenical soap for preserving skins, and things that I just happened to like: for example, a doll's bureau with a secret compartment for hiding small objects. Dolls were not a part of my world.

If my parents had wished to foster my passionate interest in natural history—and they did not—they would have discovered that we lived only three miles from the Brush Hill Bird Club and would have enrolled me as a junior member. It was my great good fortune *not* to learn of the Brush Hill Bird Club. Perhaps I'm unfair, but I have a sneaking suspicion that they would have dissuaded me from my private collections, my gun, and perhaps even from my personal wild animal hospital to which our neighbors brought injured creatures for me to nurse back to health.

Among these injured creatures was a kestrel. I nursed her back to health and recognized that she was a falcon, so I read all the books I could on falconry. I trained her until she took English sparrows day after day as they went to roost in the ivy on the church.

The period 1915 to 1925 was a time of transition in American ornithology. Bird clubs were forming, and for many people birding was becoming a pastime. By some fluke of fate I found the professionals, and the direction my life was to take was clearly spelled out by the time I was fifteen. Later, as an adult, I published technical papers on all the subjects that I have touched on.

This book traces the evolution of many ornithological techniques. When I got engaged to Frederick Hamerstrom in 1929, he said, "Lay off that insect stuff. We're going to work on game animals." I gave my collections to museums and moved into the wildlife profession with him. Frederick is quiet and scholarly.

I had no idea that my life with the birds would ever turn out to be so rowdy — or that I someday would borrow falconry techniques to study fighting chicadees!

A *bird in the hand* . . .

The more I watched birds, the more I realized that what I was seeing was meaningless because I could not distinguish between individuals. Two good chickadee fights outside the kitchen window of our farmhouse in the winter of 1935–1936 exasperated me to the point of taking action. Was one bully responsible for the fights? Were the chickadees behaving like children in school? I devised a trap by putting an upside-down box of wire over a chunk of suet, propping it up with a stick to which I fastened a string that ran into the kitchen. Catching that bully was no problem. My problem was finding a way to mark it. My first bird band was of solid gold. I dumped my jewelry box on the kitchen table, found a necklace that I could pry apart, and bird #1 carried a gold band. I released it, and it went straight back to feast on the chunk of suet inside the trap. Another chickadee zoomed down to feed; #1 flew to a jack pine, and with a mighty jerk I pulled the string again. Happily I prepared to destroy more jewelry, and holding the chickadee in one hand, I pushed useless rings, bracelets and necklaces to one side of the table hoping to unearth a band. Only another link of gold necklace would fit a chickadee. But I had forgotten which leg I had put my first band on, so I couldn't name them Gold-right and Gold-left!

If I had known how easy it was going to be to catch chickadees, I would simply have let the bird go. Instead, I put it in a paper bag, closed the top with a piece of string, and punched a few air holes in the paper with a darning needle. I put on a jacket and made a hasty trip to the dump back of our barn. Dumps always give me ideas.

I wandered—not aimlessly—on the dump and stopped to reflect when I found some broken celluloid toys. It was cold and snow sputtered around me, but I managed to light a match and learned that celluloid could be bent when heated. When Frederick came home he found me, not cooking supper, but emptying my pockets of headless dolls, toy wagons, and a broken fire engine. "Aren't the colors wonderful?"

6

"Colors?"

"Yes, for chickadee bands. Can you find me a nail the size of a chickadee's leg?"

"You let the fire die down!"

And it was getting dark. Guiltily, I quickly heated up a pot of mulligan while Frederick, who had had a long day in the field, got our wood stove roaring and found me a nail the size of a chickadee's leg.

After eating and warming up, we spent an interesting evening together. We cut the toys apart into narrow strips. One by one we dipped the strips into boiling water; wrapped them around the nail; held them in the boiling water bath, using two pairs of pliers (and two people). Then we placed each neatly coiled band on a dish towel. The chickadee from the paper bag bit persistently as we struggled to put his band on. Somewhat reluctantly, I released him into the night, but the next morning Red-right was one of the first birds at the feeder.

Sometimes I thought I caught a glimpse of a gold band, but I couldn't be sure. "There has got to be a better way of marking chickadees," I grumbled.

Frederick, who was shaving by the kitchen window near the feeder and trying to spot bands for me, heartily agreed. "How about colored feathers?"

I found a useful feather in the speculum of a dead mallard. By cutting off most of one feather in a chickadee's tail and just the tip of the shaft of the mallard feather, I could slip one over the other and glue the shafts together with household cement. The mallard feather was too big so I trimmed it down to size with Frederick's mustache-cutting scissors. Mallard Speculum could be spotted instantly at about ten yards without binoculars and none of the other chickadees seemed to pay the slightest attention to his strange plumage.

Blue Jay Tail followed, but I had to use a new technique: I glued a morsel of a blue jay's feather directly onto a chickadee tail feather. Woodpeckers and a guinea fowl provided further useful feathers, but it was plain that we were dealing with a large flock and needed more colors quickly.

A trip to the university at Madison was a bonanza. Dear old Professor Wagner lent me funnel traps and Biological Survey bird bands, but best of all the curator of the museum actually pulled some brightly colored feathers out of his bird skins for me, and then gave me a whole (badly stuffed) parrot gorgeously gaudy with vivid

7

primary colors. When winter came again I was ready to start my study.

Here is what I found out:

1. It takes more than one year of winter feeding to build up a chickadee flock.
2. Big flocks precipitate fighting.
3. A chickadee, once worsted by a given individual, could count on being vanquished by that same bird at the next enounter. We saw only one exception in seventy-six fights.
4. Chickadees bite, and have an uncanny knack for seizing upon the tender skin at the base of a fingernail.

At the time Frederick and I were both students of Aldo Leopold and we had planned to use part of the prairie chicken research for my master's thesis, but both Leopold and I saw that no part of the chicken work split off handily. It needed to fit together and it only made sense for the whole kit and kaboodle to go into one thesis—Frederick's.

Finally I said, "Aldo, I have some material on chickadees."

"*Chickadees,* Fran?" He settled back in his chair. "I had no idea that you were interested in chickadees."

"I've been watching color-marked populations for three winters."

"*Color-marked!*"

A couple of weeks later I brought him my master's thesis almost in final form: "Dominance in Winter Flocks of Chickadees," which was published in the *Wilson Bulletin* 54(1942):32–42.

Leopold asked, "How did you come on this idea?"

I didn't think to tell him of a child who had perched in an ancient oak, watching unmarked birds coming to drink at a high ephemeral pool.

History of the bal-chatri

Chickadees were a stepping stone—no more. I like the little creatures, but birds of prey have cast their spell on me.

My early attempts to catch hawks and owls met with little success. I tried nets and I tried horsehair nooses attached to carrion. I suppose I caught roughly one out of every 600 hawks I tried for.

Then Dan Berger and Helmut Mueller, both of the Cedar Grove Ornithological Station, learned of an ancient trapping device, used in India to catch accipiters. (These sprint-flying long-tailed hawks normally catch smaller birds on the wing.) The trap was called a bal-chatri or boy's umbrella.

The bal-chatri was like an upside-down basket. All that it took to make one was some string and some strips of cane, which projected so that they could be pushed into the ground. A live lure bird fastened inside attracted the hawk. The trap was covered with horse hair nooses to catch the hawk by its feet.

At first Berger and Mueller were rather secretive about their discovery, but finally they walked into our kitchen and showed me a disk-like container made of hardware cloth.

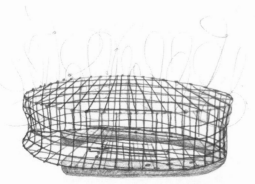

The bottom was weighted with a horseshoe, leaving room for a door just big enough to push a mouse inside. The top was covered with four-pound test monofilament nooses!

9

They took me trapping. Berger—moving steadily at about twenty miles per hour—drove past a kestrel sitting on a phone pole. Just after we passed the kestrel, Mueller pitched the baited bal-chatri out of the open door. The hum of the motor remained even as we continued down the road and, before we had a chance to turn around, the kestrel was on the trap!

"Caught!" I yelled. "Let's go."

Berger calmly parked the car by the roadside, and he and Mueller watched the action with binoculars. The kestrel, not caught at all, left the trap and hovered about three feet above it. "Move," Helmut muttered. He was talking to the mouse.

The kestrel swooped onto the trap and made repeated attempts to foot the mouse. Next it rested, and when it decided to fly off, it dragged the trap with it. Really caught.

After we had processed and banded the bird, Helmut said, "We have caught quite a few kestrels this way." He added kindly, "Maybe you can catch some too." (I caught seventy-two in the next two years.)

My first bal-chatri was made of a deep-fat fryer. I left the handle on the fryer so I could scoop the trap up more easily from a moving car. (This was probably the first of many idiotic "improvements" on the bal-chatri.)

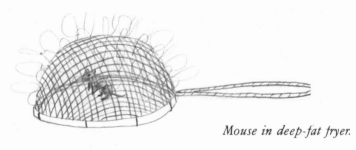

Mouse in deep-fat fryer.

After monkeying around trying to fashion nooses out of monofilament (a relatively new material in 1955), I pulled out my supply of horsehair, clipped from horses' tails—normally with the permission of the owners—and covered my invention with horsehair nooses because I found them quicker to make. For kestrels, I used about four hairs. I held the ends of the strands in both hands and twisted in opposite directions until the four-strand thread was so tight that it started to coil on itself. Then I seized the middle of the thread in my teeth, brought the ends together and directed them into a smooth cord with a loop at the end. Presto! A noose was made.

Noose made of horsehair.

Mueller gave me a mouse. It was enchanted. One by one it clipped off all my horsehair nooses and made itself a nest.

I learned to use monofilament. Only certain rather uncommon individual house mice are such confirmed homebodies that they build nests out of monofilament nooses.

In a matter of weeks the news spread, and Wisconsin and Illinois raptor enthusiasts feverishly started making bal-chatris and catching adult hawks. We discovered that it was hard to obtain live house mice. (Movie theatres, near the popcorn stand, were good places to pounce on unwary individuals.)

Using mice in bal-chatris had a curious byproduct: hawk trappers learned to give each other those unusual wretched mice that like to build nests out of monofilament.

The modernization of the bal-chatri was a major breakthrough in raptor research. 'The Bal-Chatri: A Trap for the Birds of Prey," by Daniel D. Berger and Helmut C. Mueller (*Bird-Banding* 30(1959): 18–26) finally facilitated the capture of adults on their breeding grounds. Before the appeaarance of this paper most raptors banded were nestlings or migrants, so intensive ecological research had not been possible.

The original disk-shaped trap tended to roll
into ditches, so most of us use
quonset-shaped bal-chatris nowadays.

Hawk owl expedition: I select personnel

It was Christmas cards that alerted us to the hawk owl invasion of 1962. Those little personal notes at the bottom of cards mentioned hawk owls in Ontario, in Michigan, and in Minnesota. Frederick and I moved our vacation to winter. Helmut and Nancy Mueller and Dan Berger agreed that this was a once-in-a-lifetime opportunity, and a hawk owl banding trip was in order. I arranged for a Canadian banding permit and Errol Schluter, who wanted to come, offered to make special hawk owl traps. "Errol," Frederick stated, "only one car is going and we just don't have room for you."

Errol's disappointment was ill-concealed. But the personnel of the expedition underwent a swift change. Our supervisor in the Wisconsin Conservation Department denied Frederick's winter vacation because of a special report deadline. Helmut telephoned that he and Nancy and Dan couldn't come. In deep distress I watched all our plans fall apart, but my eye trouble was to save the day. My right eye flared up from time to time, and each time that it got to looking like raspberry jam, pain drove me to the doctor.

Dr. Anderson gave me a lengthy examination and said, "I want you to put a hot pack on that eye every four hours." And then he added the clincher: "And I don't want you to read for three weeks."

I could have kissed him. I couldn't read, so I was free to go. "Schluter can come. I'll call him."

Frederick asked gently, "What's he like?"

I shrugged. "You know Errol Schluter: he trapped snowy owls."

"Yes," Frederick expostulated, "but do you know more about him?"

"He can catch hawks too."

Then I phoned Errol, who could hardly believe his good fortune. "I'll be there in four hours."

The Hawk Owl Expedition had a certain feverish quality to it. Berger, after having said he couldn't come, had second thoughts and sent a postcard stating: Call me collect if catching is good.

Schluter and I were on our way by midmorning with the bus packed with many mice, some starlings and pigeons, and a half-grown rat, and provisions for all the vertebrates—including us.

We headed for Ontario, by way of the Upper Peninsula of Michigan, in high spirits and with very little money. We baited up a couple of traps as soon as we crossed the Wisconsin-Michigan line and peered at every treetop ready to pitch out a trap at a moment's notice. A clean washrag and a quart thermos bottle of hot water, which I replenished every time we lit up the Coleman stove to cook a meal, solved the little problem of hot packs, obediently held to my eye every four hours.

The country got lonelier and lonelier. Sleeping by the wayside rolled in sleeping bags was not appealing. At deep dusk I studied the Michigan map for a moment and announced, "I know where we can spend the night *free!* Cusino Ranger Station. It has heat and bunks and an indoor kitchen. And it has bath tubs with hot water!"

Cold had settled between our shoulder blades. Cold and the strain of driving on slippery roads and excitement had worn me out. Sleep!

The ranger said, "You've got the place to yourselves. Stay as long as you like."

I selected a room with a private bath and Errol chose a big room with a lot of bunks. We left the sleeping bags in the bus and made our beds with clean white sheets from the linen shelves. A sign admonished us to put soiled linen down the laundry chute.

I cooked supper, cleaned up, and said, "Why don't you set the alarm for ten past six, cook breakfast, and then call me?"

My room was warm and so was the bathroom. I filled the tub with luxurious hot water and was about to step in when Errol knocked on my door. "Fran, I hear something outside."

"Just a minute." I scrambled back into my clothes and we rushed out. Errol described the sound. We strained our ears in the frosty night, but the sound was not repeated.

By now the water in the tub was lukewarm, so I let some out and replenished my bath. Then I soaked for a long time before crawling between smooth white sheets under Hudson Bay blankets.

I missed Frederick.

Errol knocked again. "Fran, there's something I want to show you!"

"I'll get dressed."

Errol wanted to show me some photographs he'd found on display. He thought they were wonderful. I considered them perfectly standard conservation department shots—of bucks, ducks, and beavers—just run-of-the-mill, but it didn't seem kind to dampen his enthusiasm. At length, I said, "Don't call me till breakfast is ready."

Just looking at all those stacks of pictures had tested my politeness almost beyond endurance. I had another hot bath for good measure and to recover from the strain.

The next time that Errol knocked I could smell bacon cooking. I had a quick cold bath, scrambled into my clothes, dismantled the bed and shoved the sheets and pillow case down the laundry chute.

Breakfast cooked by Errol was excellent, but he is a deliberate man with a knife and fork. Finally I tried to hurry him. "Isn't it time we get the show on the road?"

Errol looked at his watch. He shook it and looked at it again. He jumped up in dismay to look at the alarm clock.

"Er . . ." Errol cleared his throat. "I must have set the clock wrong . . . it's 2:30."

I set the clock for 6:10 and handed it to him, muttering, "Back to bed."

The mattress on my bed lay bare with the Hudson Bay blankets neatly folded at the foot, just the way I'd found them.

And my sheets were down the laundry chute.

This time I just pulled off my boots, shook open the blankets and curled up under their fluffy warmth. Sleep, oh, blessed sleep.

At 6:30 Errol knocked again, and again I smelled bacon. It seemed to me that we had just eaten. I gobbled the bacon, drank a cup of coffee, and got our gear into the bus while Errol concentrated on large mouthfuls of food. He ate his own huge breakfast and most of the breakfast he'd fixed for me. I intercepted just enough scrambled eggs to put between two slices of bread to save for later.

We made a good start. Thanks to new snow tires we were able to get the car out onto the main road. Drifts piled high. The snow plows were busy, and more and more blinding snow swept almost horizontally across the windshield.

Errol kept driving down the middle of the highway—right where that white line probably lay under the snow. It is my policy not to tell people how to drive automobiles, but as we could see only a few yards through the blizzard, I finally said, "Stay on the right side of the road, or somebody'll smash into us!"

Then I went aft to stretch out on the back seat and catch up on some of the sleep I'd missed last night.

Karrump! the car came to a sickening sudden stop and I was pitched onto the floor wide awake.

The car was buried in a drift and partly down a ditch on the *left* side of the road. We shoveled, cutting big snow blocks and heaving them aside. After about an hour, two small French Canadians stopped to help us. (No other cars had passed.) They didn't cut blocks of snow. They wielded small shovels with a kind of frenzy whizzing the drift away like a snow blower, until they could get us out. For payment they accepted the "egg salad" sandwich.

It was my turn to drive. We poked through the blizzard and Errol didn't waste a moment. Our chances of seeing a hawk owl struck me as absolutely nil. But he watched both sides of the road, and one hand was always ready to throw out a baited trap at a moment's notice. Harriet, our half-grown rat, very sensibly dozed in the bal-chatri, but Schluter never relaxed his efforts.

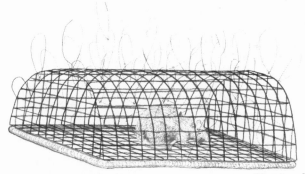

Harriet dozing.

"What time is it?" I asked after a while.

"10:30."

We had traveled together for twenty-four hours.

If Frederick asked me now what Schluter was like I could tell him a great deal more . . .

15

Quite *all right, thank you*

Hawk owls are diurnal and are said to spend much of their time perched on the very tip tops of spruce trees. South of Sudbury, Ontario, we drove down a broad, lonely, well-plowed highway and tried to examine the top of every tree. The blizzard was behind us and millions of spruces, tall and unharvested, exposed their tips against the bright blue sky. It was as though the world's supply of Christmas trees grew on either side of the road. It was the second day of our expedition. Our hopes were high. We cruised at about thirty miles an hour, and each of us devoted attention to his side of the road only. There was very little bird life of any sort.

Suddenly I said, "This is a funny place to find a pigeon."

We skidded to a stop and the "pigeon" — in an unpigeonlike manner — swooped to perch atop a spruce at the side of a steep ravine.

Our first hawk owl was in sight!

We both knew that any creature is more apt to be spooked by two people rather than one. I handed Errol three traps: a big bal-chatri baited with a pigeon, a medium-sized one baited with Harriet (not an easy one to handle, because Harriet had a knack for biting fingers through the wire mesh of her container), and a small heavy trap baited with one brown mouse.

Errol clambered up the hard massive ridge thrown up by the plows, stepped confidently beyond and fell into three feet of virgin snow. He righted himself, found the traps, and managed to take almost three steps.

"Hold on!" I called.

I tossed Errol my snowshoes. He fumbled with the harnesses until I was sure the hawk owl would take off from sheer boredom.

Finally in desperation, I jumped into the snow next to him, got him to bend his knees so I could tighten the straps and passed him the traps.

Errol took two steps, stepped on the edge of his own snowshoe and fell down again.

"Sorry," he mumbled.

"Walk down the middle of the ravine," I admonished, "and don't stop. Walk right past the owl and *keep going!*"

"I know," Errol spoke in his usual deliberate manner. "You just want me to amble by."

Errol was pretty new at trapping, but he caught on quickly. I didn't learn until later that he had never been on snowshoes before. He walked in a very strange manner—keeping his feet as far apart as possible, instead of swinging ahead with long paces. It was agonizing to watch his progress down the ravine through binoculars.

Errol set the pigeon on the snow, just peeking sidewise at the owl to see if it was in position, took a few paces more and dropped the trap with the mouse. This heavy little trap disappeared under three feet of fluffy snow. Errol tried to take a step backwards to retrieve the mouse. The tail of his snowshoe slid deep down behind him and he fell again.

For a long time I couldn't see what was happening. He didn't seem to want to get up.

The hawk owl kept turning its head as though it wanted to go somewhere else, but it stayed, sitting atop its spruce.

Finally Errol righted himself and moved on down the ravine with only Harriet to deposit. I noted with approval that he set her near some emergent vegetation where it seemed a rat might be.

At last he lumbered part way up the side of the ravine and hid behind a clump of young spruces.

When I looked at the spruce top again the owl was gone. It was working a trap!

The toes of owls are built just wrong to get caught in nooses easily. All of that "fur" keeps any noose from closing properly and encourages it to slide off. I could watch the little owl fight and struggle with beating wings to get at the bait. It paused for a moment and then struggled and fought its way to the top of the trap.

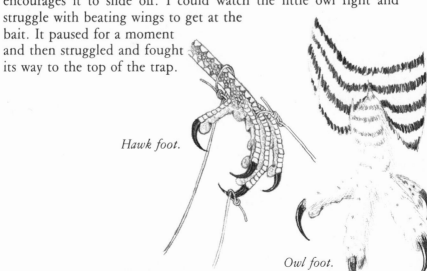

Hawk foot.

Owl foot.

Errol stayed crouched in hiding. We both knew that the longer the owl worked the trap, the better its chances of getting caught. Then the owl gave the signal! It tried to fly away dragging the trap with it.

I yelled, "Run!"

Nobody heard me because Errol was well out of earshot. He started down the side of the ravine carefully placing each snowshoe in position at every step. I tried to run toward the owl, but went in over my knees—even on the snowshoe trail. It was hopeless, and we had only one pair of snowshoes with us.

Like a slow motion movie, Errol moved over to the owl—nearer and nearer—and then he hand-grabbed it. He struggled back to the bus holding the owl and the trap in front of him as though he were carrying a silver tray with cups full of tea.

Remembering Dr. Anderson's admonition, I poured hot water out of the thermos and quickly soaked my eye. I could hardly wait to see my first hawk owl, but there was no point in wasting time. At last, Errol handed me the owl.

That exquisite little bird was prepared to fight my fingers to the death as I gently struggled to work off the four nooses that held it by its formidable feet. And it bit successfully and ferociously. Its legs were so short that every time I got my fingers close enough to pull a noose free it dug into my bare hands with its beak. I was outraged! Very few birds of prey bite. Kestrels sometimes do and bald eagles have an unnerving way of taking a nip, but I'd never had trouble with an owl before.

Errol said, "You're bleeding."

The owl bit another piece of my tender skin as I slipped off the first noose.

"Do you want me to hold its head?"

"Dammit, NO!" I wasn't going to let a bird that size fake me out. Blood trickled off my fingers and dripped in the snow, and it was with a sigh that I finally stuffed the owl into a tube to process it.

WEIGHT: *353 grams*
WING CHORD: *230 millimeters*
FLATTENED WING: *237 millimeters*
TAIL: *165 millimeters*
SEX: *Unknown*
DATE: *Jan. 13, 1963*
LOCATION:

Location? We didn't know where we were! No traffic had passed. There was no one to ask, and I hadn't noted the mileage at the last town.

Errol put on the snowshoes and went to pick up the remaining sets and I took the owl out of the tube to admire it.

And then I saw a car coming. It was a police car. What luck!

I jumped out of the bus and flagged the police down with the owl which I just happened to be holding. My black eye patch and tattered parka didn't make me look especially respectable. I leaned inside the window they had opened and asked, "Where am I?"

"Why, you're on Highway 69 heading toward Parry Sound."

"I need to know exactly where I am. How far from the nearest town."

"Are you in trouble?"

"Trouble? No. We just caught an owl. This one." I flicked it up so they could see it and it bit me again.

"It's alive!" one of them murmured.

"Yes, it's alive. I'm going to band it and I need to know exactly where we are."

"Where did you catch it?"

"Here. Don't you have a map? Can't you tell me exactly where we are?"

I put the owl's head in my pocket so it would quit biting me and peered at the map they produced. "We are about two and a half miles north of the Wahnapitae River. Is that close enough?"

"That's fine. Thank you very much."

The driver seemed reluctant to leave. I like to think I look like a pirate with my black patch, but sometimes I don't quite bring it off and look purely pitiful. Errol's gait on snowshoes had improved, but every step required concentration. His face was set and he looked purely pitiful too, clambering up the ravine with a trap in each hand.

They looked first at him and then at me.

"Ma'am, are you sure you are all right?"

I adjusted my patch, which kept slipping, put on my Queen Victoria smile and answered, "*Quite* all right, thank you."

Need another hawk owl?

"Fran, how do you go backwards on snowshoes?"

"*Backwards?*"

I took my eyes off the spruce tips to look at Errol's face. The question was serious.

"You don't. You just run around in a little circle until you get back to where you want to be."

"I wish I'd known that before."

We drove a mile or two in silence and then Errol spoke again.

"I had a little trouble back there getting out the traps."

"I saw you fall down."

"Well, what really happened is that when I put out that heavy little bal-chatri, it went clean out of sight with the mouse."

"Yes?"

"And then I had to back up, but I sat down instead."

"So I saw." My answer was dry.

"And then I had to build up a little snowpile that was solid enough so the owl could see the mouse."

"Oh!"

For a moment I tried to visualize Errol, both feet imprisoned by snowshoes, finding the trap by feel, and then manufacturing a snow-pile high enough so the owl could see the mouse.

"Under the circumstances I think you did it all very quickly."

The rest of the day is blurred in my memory. We caught six more hawk owls before dark, and then we pulled into a town where I begged empty cardboard boxes to store the owls in until I could get to my bands. For comfort each owl needed an individual box.

For some mysterious reason the Canadian Banding Office had sent all my bands to Toronto. Actually it was my own fault. The office had asked for my address and instead of giving them General Delivery at my port of entry, I gave them Bill Gunn's home address thinking it sounded respectably stylish. So now we had to keep the owls in storage until we got to Bill's.

As Errol and I carried empty boxes out of the store, he grinned delightedly and said, "We had good catching!"

Good catching. I ran back into the store and phoned Dan Berger collect.

Berger was ready. "Pick me up at the Toronto airport at 11:15, Flight 327."

We dashed to the airport and now a third member was added to the expedition. Berger threw his luggage into the back of the bus ignoring the piles of boxes, and climbed into the front seat with us.

The ride to the Gunns was uneventful until Berger asked, "How many did you catch?"

"Seven."

He sighed, "I wish I'd seen them!"

I gave Errol a sharp kick and changed the subject immediately.

The hawk owls made melodious music in their boxes, drowning out the cooing of pigeons and the songs of starlings. Berger's only comment was, "It seems you've got quite a singer starling."

When we reached the Gunns, I announced, "If you really want to see those hawk owls, help carry them into the house."

Berger gasped, "Heavens to Betsy!"

We banded them and processed them and admired them, and Bill got out his fabulous sound equipment and recorded their calls— first singing in their boxes—and then we produced a stuffed horned owl and Bill recorded their rather violent objection to such a sight, in his living room. At last we went to bed.

Berger is very well organized. When Errol and I got up Berger

21

had already telephoned a variety of knowledgeable ornithologists to learn where we might find more hawk owls.

The nearest one was in the city park! But Berger, who was talking with someone in authority, was having difficulties.

"We don't just want to see it. We want to band it . . .

"We won't hurt it. We'll just weigh it and measure it and let it go right there . . .

"We have no intention of scaring it away. We'll even give you another hawk owl . . .

"You can all watch. We'll meet you there at 10:30."

Berger put down the receiver, sighed, and said, "Let's go before they change their minds."

Birders, swaddled in fur parkas, were already gathered near a hawk owl perched on the top of an *Osage orange!* (Amazing that a bird from so far north would ever encounter this southern vegetation.)

We used a small bal-chatri, baited with one brown mouse, and caught it in less than two minutes. After processing we lined up for a photograph (not easy with bare hands and biting birds).

Seven hawk owls and, left to right, Bill Gunn, Errol Schluter, Fran Hamerstrom, and Dan Berger.

The leader of the group said, "You promised us another hawk owl."

Berger simply commanded, "Release," and eight hawk owls flew to nearby perches as though to adopt this very part of the park for their new home.

What are you doing?

Berger laid out our route.

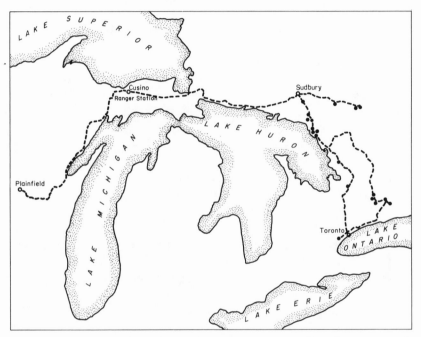

Hawk Owl Expedition. Over 2,000 miles of winter travel.

We saw twenty-three hawk owls and caught twenty-one. They were incredibly tame. One in an apple tree almost let us grab it. Tame—and ferocious—footing and biting with vigor as soon as we had them in hand.

Keeping expenses down was a big problem. Berger arranged to have us spend the night in the basement of Gus Yaki's laundromat. Laundromat-and-bird-people all seem to know one another and Berger and Gus were both. Gus showed us his exotic cage birds, guided us to a hawk owl (promptly caught), and then took us to his laundromat to spend the night. I feared the basement might be a bit hot for sleeping. It was worse. A pipe had burst and the floor was covered with inches of steaming water, so we slept in Gus's house.

Then we moved on to our next contact: a woodsman who knew of two hawk owls. He showed us the first hawk owl that paid no attention to any of our traps. It just perched on the tip top of a spruce and viewed the scenery. And it was getting late in the day.

"How far is the other hawk owl?" Berger asked anxiously.

"About thirty or forty miles," our guide answered.

"Fran can stay."

Our guide gasped. "You mean you're going to leave the lady here *alone?*"

"What's wrong with here?"

"Here" was an opening, a spruce bog, a farmhouse in the distance, and a lazy-daisy hawk owl. Berger handed me two traps, one baited with a pigeon; Harriet, our half-grown rat, was in the other.

"Take good care of Harriet."

I put out the sets in sight of the owl and buried myself in a snow drift to keep warm, and to worry. I had two worries and both increased as the day wore on. I feared the owl might fly away and never get caught. It kept looking around as though wishing to start out on a journey. And I feared for Harriet: her bare, pinkish feet and her long naked tail were not well protected from frost-bite.

Pigeons are tough and long hours at subzero temperatures are no problem for them. Rats do well if they can build themselves a nest of hay or rags, but a nest would hide her and defeat her present mission.

Each time I decided to rescue Harriet, the owl seemed especially alert. I argued with myself:

I could just leave the pigeon out. Hawk owls take ptarmigan in the wild—and a pigeon is much smaller.

But maybe this is a male owl and afraid of pigeons?

I wish I had a watch. I could leave Harriet out for ten more minutes and then pick her up no matter what.

My face isn't getting frost-bitten.

But it's out of the wind.

Harriet is right on top of the snow in the . . .

then the hawk owl hit.

Caught!

When the VW bus finally reappeared I was crouched by the

roadside. The pigeon dozed in his trap, the hawk owl reposed in a nylon stocking and Harriet sat on my lap, *inside* my parka crunching a frozen carrot.

I was about to learn what it is like to be really cold. After another day's catching, we got caught in a blizzard—unable to see to move the car at all. Berger slept on the back seat, Errol on the front seat, and I curled up on the floor. We had a good meal of split pea soup before going to bed and Berger alerted us to another catching possibility. We might spot a great gray owl!

That night I daydreamed of catching my first great gray owl and, as the cold ate down between my shoulder blades, of sleeping in steaming water on a laundromat floor.

At length it was daybreak and the motion of the car woke me up. I stayed curled up on the floor soothed by the hum of the motor.

Suddenly somebody shouted, "Great gray!"

The side door of the bus slid open with a bang. I reached for a baited trap and was jerked back to the floor. I slid upward, grabbed the trap and hurled it out onto the snow.

Errol said, "Sorry, just a stump. Looked like an owl."

Wide awake now, I struggled to get up, mystified by my difficulties. My parka and sleeping bag were firmly frozen to the floor.

Much as we were determined to save money, we spent the next night in a hotel. We looked for a cheap hotel, and we looked forward to hot baths. Berger, who is used to city driving, took a corner too fast, there was a crash, and the stench of mice dominated the atmosphere. The mouse container lay upside down on the floor and brown mice and white mice scurried in every direction, scooting through spilled apple cores, hay, and sawdust. Fortunately most of them were demoralized by having their world scattered on the floor.

We plunged into action. Errol caught three mice in one swoop and held them cupped in his hands while I righted the container so he could dump them back home. Berger seized a large white buck which promptly bit him.

"Be gentle with my mice," I yelled, "they only bite if you're rough."

He examined his bleeding index finger, while Errol and I continued to grab mice and plop them into their container. Berger, who is absolutely stoic when a big red-tailed hawk foots him, objects to being bitten by mice. Gamely he made a try for another mouse, trying to sweep it up as one grabs a fly from a table top. "Look," I pleaded, "you're teaching my mice to bite. Just cup your hands, or grab them by the tail."

His answer was, "Let's find that hotel and eat."

At last we were parked in a seedy alley back of an ancient hotel. Trash cans and rubbish dumped on the snow made this a far from scenic spot, but there was a bright light by the back door that would give our bait birds a chance to feast all night if they wanted to.

A small gray mouse sat on top of a shiny black suitcase, polishing its whiskers.

"You get it," Errol whispered.

I cupped my hands and slid them slowly, slowly upward, grabbed, and missed. "Damn! Well, let's set the live traps for the rest."

Errol opened our new jar of peanut butter and we set to baiting traps. Just as I was smearing a trail of peanut butter on the floor of the car to lead some unwary mouse to my trap, a bright light pierced the gloom inside the vehicle and someone knocked on the door. Berger opened it.

The police.

"Excuse me, Sir, what are you doing?"

"We are trapping mice."

Errol finished patting a dab of peanut butter next to a blue sleeping bag on the floor. For a moment I considered wiping my buttered hands on my trousers. Instead I licked my fingers.

The courtesy of Canadian police officers is impressive. Here were two rugged officers peering into our messy vehicle. One said, "Thank you, Sir, we are sorry to have disturbed you."

On keeping the working mouse happy

Aldo Leopold once said, "Wildlife research is the greatest sport there is." I agree wholeheartedly and the myriad miserable little details connected with mice have not changed my mind.

We have tried a variety of species for hawk trapping. Meadow mice, deer mice, and jumping mice are second rate as lures; they spend too much time sitting perfectly still. The common house mouse is a restless creature, but it is so agile that it escapes any time it has a chance. Ordinary laboratory mice combine restlessness with reasonable docility and an ability to learn what we want them to learn. They like to go *down,* so it is easy to persuade them to return to their containers. I open the door of the trap and let them find their way downwards to rejoin their friends. They don't like to be blown at. Talking to a mouse is useless, but blowing in their faces steers them, and gives one a sense of power.

If one is traveling with more than one container of mice, it pays to put each mouse back where it belongs: among its *friends.* A mouse, placed among strangers, is at a strong disadvantage. It often has to fight its neighbors to survive. The squealing and scuffling is distressful to the human ear, and the fight may result in severe injury or death before one can manage to rescue the innocent victim.

Upon occasion it is necessary to add new mice to a container. The trick is to scrub the container thoroughly, getting rid of the nice homey smell that seems to give mice their sense of well-being *and* aggression toward strangers. A newly scrubbed container with fresh litter puts all the mice one cares to dump into it at an equal disadvantage, and everybody settles down with a minimum of fights — almost always. Some really feisty individuals seem possessed with a strong sense of territoriality and go after any individual that tries to run away. A few empty tin cans solve this problem. The weak and the timid take refuge in these and may hole up for over an hour, by which time their body scent has probably permeated the vicinity and — at last — they have become acceptable.

which time their body scent has probably permeated the vicinity and—at last—they have become acceptable.

I thought we had this all worked out until we went trapping with Dr. Bill Foreyt on his wedding day. He had managed to catch a wild house mouse and his bride wanted to put it into our big mouse container. "No," I objected, "You'll never catch it in there and besides our mice will kill it in nothing flat!"

With some difficulty (Bill normally handles cattle), he coerced the wild mouse into a bal-chatri and succeeded in catching his first hawk—a kestrel. At dusk, Bill tried to transfer his mouse into a coffee can so he could take it home. The mouse, having no intention of slithering down into a shiny, empty coffee can, leapt for freedom and scooted behind our food box.

Frederick sighed. He does not enjoy orange seeds in his boots, socks converted to mouse nests, and having to examine each cup to see if a mouse had been there first. It reminded me of the early days before we discovered the merits of lab mice and when we almost always had at least one mouse loose in the car. I sighed too, but went to bed untroubled.

The next morning I checked the snap trap. It wasn't even sprung. Then I took a carrot and some apple cores out of my pocket to feed our fine group of lab mice some goodies. The wild mouse had squeezed his way into the container and jumped for freedom the moment I lifted the lid.

Slowly I put the carrot and the two apples cores back into my pocket and gently stirred the litter. Ten lab mice were dead; murdered, directly or indirectly, by one small invader.

Mice can make a bloody mess of things. Newborn mice sometimes simply disappear—devoured—so we set up maternity wards for pregnant females where they can rear their young in peace. But cannibalism is not uncommon. (I have no idea what starts it and the only way I know to stop it is to kill every cannibal pronto.) Once it gets started it seems to be contagious as though the taste of fresh blood had something to do with it. When we first started raising mice they had a constant supply of water, and it seems to me that cannibalism occurred more frequently in those days. At length we discovered that water caused the mice to be sluggish, lacking in stamina, and far more stinky. We have now raised many, many generations of mice without water. We substitute succulence: lettuce, zucchini, tomatoes, cabbage, dandelions, plantain, apple cores, pears, peaches, grapes, mangoes, papaya, but *not* bananas!

I have always been fond of slightly over-ripe bananas. Somehow the soft, honey-colored portions of a banana with its concentrated sugar content is like an extravagant candy.

One evening in winter a shipment of fifty mice arrived from the university, well furnished with commercial mouse food, but without succulence — and five slightly over-ripe bananas were all that I could find. They were *my* bananas and I didn't like to give them up, but in this emergency I made an almost supreme sacrifice for the new mice. I peeled each banana, bit out the yummy little brownish areas for *my* pleasure and gave the mice the rest.

That night I had a stomach ache — an occurrence so rare as to be noteworthy.

The next morning every mouse was dead.

The electric mouse

Barred owls had set up a winter territory in the woods north of our house. After working in the field all day trapping and banding prairie chickens, I didn't wish to spend my evenings crouched in the snow waiting for these owls to finally discover a mouse in a balchatri. Furthermore, I had no way of keeping the mouse warm. Since any one mouse could not survive our winter cold for more than about twenty minutes, a continuous relay of new mice would be necessary.

I wanted something better than a live mouse.

Even in summer live mice need attention: food, succulence, bedding, and protection from heat. Actually they are about as much trouble to look after as horses—but on a smaller scale.

It took me some time to plan my electric mouse, and even longer to find the materials and manufacture this contraption. I needed:
- a battery-driven motor
- a wire dome, covered with nooses
- a lively-looking stuffed mouse
- a wire, at the end of which the mouse would go round and round, propelled by the motor.

And first, I needed the motor. Luck was with me. Even before trying to shop for one, I stopped in a small country tavern where a gaudy whiskey ad drew my immediate attention. It was a large cardboard poster depicting a cowboy on a rearing horse. His right arm was held aloft, and a battery-driven motor activated a lasso that went round and round.

"Could I buy that?"

The owner moved around the counter to admire the picture. "Pretty, ain't it?"

It is beyond me to say that something is pretty when I don't find it so. I mumbled, "Ingenious."

"Huh? What do you want it for?"

"For catching owls."

"Huh?"

"What I want to do is to trap a mouse and stuff it. Then I'll straighten out the lasso wire and spear the mouse on the end of it. I'll have to get rid of the cardboard," I added somewhat dubiously.

He nodded enthusiastically.

"Then I'll seat the motor so the mouse goes round and round just above the ground . . . and I'll put a cage covered with nooses over the whole thing."

"Lady, you can have that ad for nothing. Want a beer?"

"Oh, thank you! . . . and I'm ever so sorry: I don't like beer."

I rushed home to get busy. I set eight snap traps in our kitchen and caught a mouse before I had finished demolishing the cowboy and testing the motor. I skinned and stuffed the mouse, but it was too heavy. So I set the oven at 200° to get it thoroughly dry and as light as possible.

Alan, just home from school, opened the oven door, and peered at the stuffed mouse on a pie plate.

"Why don't you make cookies?"

"Because I'm making an electric mouse."

With twelve-year-old wisdom, he said, "I don't think it's going to work."

"I'll bet I catch a barred owl with it before Sunday morning."

"Bet you a quarter you don't."

The mouse moved round and round over the living room floor, but it didn't look lively. I cut off most of its tail and attached a slender pliant feather to the stub. The feather swam behind the mouse like a fish not wanting to get caught. Irresistible!

"Still want to bet?"

Alan nodded and we shook hands on a quarter.

The noose dome was about a yard square and not nearly as much fun to manufacture.

Alan won the bet. The electric mouse slowed down and stopped outdoors at the slightest provocation. Dampness, a puff of snow, even a small leaf in its path immobilized it—and just a feather blowing in the breeze at the end of a stuffed mouse's tail isn't enough to cause a barred owl to attack.

The noose dome, which I fortunately made with care of sturdy welded wire, has paid off in other ways. Now and again it is impossible to net adult breeding harriers, but that fine noose dome placed over the young sometimes makes it possible to make the catch after all, and it has caught a golden eagle!

It never occurred to me that the electric mouse would give Alan ideas, or that it would ever cause him embarrassment.

Alan does not always buy his parents Christmas presents, but when he does his presents are almost always exactly right. Alan went shopping alone. He selected the toy department of a large department store and studied wind-up toys: Were they tough? Would they carry nooses? And would they stay active long enough?

A bear beating a drum had good-looking fake fur, but its energy just went into the drumsticks; a bright pink velour rabbit hopped well and could be dyed. He tested each toy on the floor. At length a saleswoman wearing rhinestone-rimmed glasses asked, "Can I help you?"

Alan said, "No, thank you," and continued winding up toy after toy.

After about forty minutes the saleswoman came back determined to let this persistent boy know that she knew a lot about toys.

She spoke with authority. "What age child is the gift for?"

Alan looked up mystified. "What *age?*"

"Yes." The rhinestones glittered like tinsel around her pale blue eyes. "Certain toys are particularly appropriate for certain ages."

Alan wound up a monkey that could climb a rope. The clerk readjusted her glasses on her small powdered nose, looked Alan straight in the face and added, "Not only are toys educational, but they also aid in mental development. What age is this for?"

"I don't know."

"You don't *know?*"

"I don't know. You see, it's for my mother."

The clerk gave a little moan and patted Alan's head. Her rhinestones twinkled as she shook her head in distress. "Oh dear!" she gasped. "I'm so sorry. I'm ever so sorry."

Happy Birthday

The failure of the electric mouse had certain fringe benefits. We spent a lot of time in Mel's woods north of our house testing that unsuccessful contraption. I looked more closely at the woods. That barred owl wasn't hooting down in the riverbottom where it belonged, but in Mel's upland woods! Could it find a mate there? Perhaps it actually had!

The only reasonably good owl trap in the early 1950s was the Verbail perch trap (named after Vernon Bailey). I had not the foggiest notion where the owl or owls perched, so this ingenious trap was useless to me.

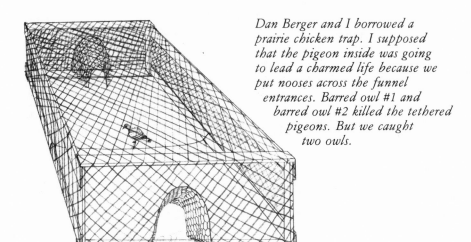

Dan Berger and I borrowed a prairie chicken trap. I supposed that the pigeon inside was going to lead a charmed life because we put nooses across the funnel entrances. Barred owl #1 and barred owl #2 killed the tethered pigeons. But we caught two owls.

Elva, then almost nine, helped weigh, measure, and band the second owl. Then she held it by its legs so I could examine its ears for parasites. "What would you like for your birthday?"

"Owls—lots of owls—all at once."

So another research project was born. I secretly hid owl #2 in the barn instead of releasing it and determined to find out if the owl in Mel's woods could attract *another* mate.

Mel liked to roam in his woods; and Mel was very fond of pigeons—not to eat.

We picked up all the feathers from the two pigeons that had been killed, so that Mel would never see signs of the massacre, and made sure that in the future bait pigeons would be safe.

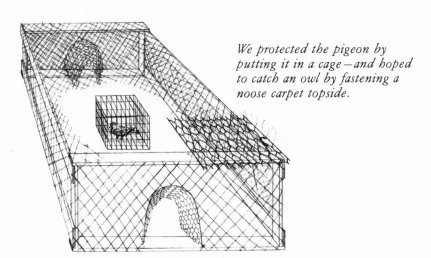

We protected the pigeon by putting it in a cage—and hoped to catch an owl by fastening a noose carpet topside.

It was mid-March. Each evening we baited the trap before supper and afterwards we drove close to the trap to watch the owl's reactions (barred owls don't seem to mind headlights). And we kept catching more owls! Each time we caught owl #1 we released it. The others went to dwell secretly in the barn.

One evening Frederick came to watch the barred owl show, and sure enough, still another owl was working our cumbersome trap.

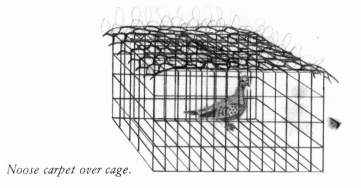

Noose carpet over cage.

"Why," he asked, "don't you do away with the prairie chicken trap? Put the nooses directly on the little pigeon cage."

I felt very stupid not to have figured out this forerunner of the big bal-chatri myself. But there was a compensation: the joy of giving a child a truly memorable birthday. When Elva came down to breakfast on March 25th there were *five* barred owls sitting on perches in the dining room. She was so happily excited that she could hardly eat, and I glowed in the realization that my daughter had a beautiful memory of her childhood to store in her heart.

Big bal-chatri ready for nooses (door is open).

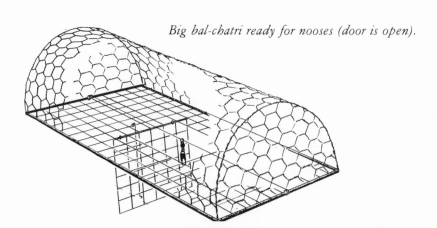

At last she asked, "Where will we let them go?"

"You choose, dear."

She chose a stream bottom woods, and we let them all go at once.

Years later, I asked her, "How old were you when we had all those owls for your birthday?"

"I don't remember."

Smiling fatuously and leaning forward to learn her answer, I asked, "But you *do* remember your birthday with all those barred owls?"

"No."

A *little trouble with the Feds*

Bird banding is conducted under a strict permit system. It is not easy to get a permit, nor is it difficult to lose one's permit. Every bird banded must be reported to the Banding Lab and, furthermore, the reports must be in a certain form—quite rightly—to adapt to computerization. I find making out my annual reports to the Fish and Wildlife Service rather like rubbing my head and patting my stomach.

When I first started banding in the 1930s, my reports went to Dr. Frederick Lincoln of the United States Biological Survey. Those were placid days. Dr. Lincoln and I occasionally corresponded about birds—a subject we both felt at home with. Eventually, the Biological Survey was replaced by the Fish and Wildlife Service and then, when computers entered the picture about 1962, I was resentful. For example, banders were informed that every bird gets its birthday on January 1. My reaction was: this is a damn lie. How can a nestling owl banded in June be a year old by January? It can't, but it is in its second *calendar* year. I was slow to grasp that this particular adaptation to the computer was not a lie, but extremely ingenious.

Perhaps because I was used to the old system, I found any change obnoxious, but now I am enormously impressed by the skill with which the Banding Lab has managed to expand its much-needed place in the bird conservation effort in North America.

As we had essentially nothing to do with other branches of the federal government, we simply referred to the Banding Lab as *the Feds*. The Feds shaped up a whole generation of bird banders, including me—and it must have taken some doing. We had to shape up or lose our permits. And then, of necessity, new restrictions limited our personal liberty. No longer could we catch any bird we pleased and color mark it. (Banders were beginning to interfere with each other's studies.) Specific permits were needed to color mark any species, and certain groups, like hummingbirds and eagles, were to be banded only by specialists. Exotic cage birds were not to carry Fish and Wildlife Service bands at all. My troubles were to come from two unexpected sources.

"Darling" (I always go to Frederick with my troubles), "I've gotten an extraordinary letter from the Feds. They accuse me of putting a red band on a roughleg."

Frederick reached for the letter. "An interesting recovery. The bird went almost due east, and taken in Canada."

"Yes, but I'm in trouble. I don't have a permit to color mark roughlegs!"

"Why did you do it then?"

"I didn't! I didn't and I wouldn't!"

"Well," Frederick has a soothing effect on me, "ask to see the letter the Canadian sent in."

So I got out the typewriter and politely requested the letter.

In due course a photocopy arrived. It was illiterate, in French, and the handwriting was execrable. It takes a slapdash type like me to be able to read such a letter without difficulty. The gist of it was, "I caught a savage bird from Washington America. I placed upon its leg a red bracelet and let it go. This red bracelet I constructed myself in order to recognize the bird again." My translation went to the Feds airmail to make my reputation lily pure.

But very shortly after, Dan Berger caught a pair of shrikes courting in the north end of the study area. Not having shrike bands with him, he brought the birds home, banded them with my bands and released them from our house. "It would be nice," he remarked, "to find out where they nest."

I thought it would be nice too.

And after about two weeks, I found their nest — at least a nest with two banded shrikes. If they were indeed the same shrikes, the information was valuable. At that time we didn't know that DDT was about to reduce Wisconsin's shrike population to a handful of pairs, but it looked as though mated shrikes could successfully be moved to nest in another area. We needed to catch them in order to be certain.

First I tried bal-chatris baited with mice — no reaction. Perhaps they had learned from Berger to distrust mice in little cages? So I substituted large grasshoppers. No luck. Maybe they didn't like little cages? I tied a good buzzy bumblebee onto the treadle of an automatic bownet (this took much patience and two pairs of forceps. Knotting a thread around the waist of a stingy bee could, I suppose, become easy with practice). Then I tried getting my sets out before daybreak hoping the shrikes would be eager for the first meal of the day.

Time was passing. The young shrikes were getting bigger and bigger, and soon they would fledge and my chances of catching the adults would be gone!

Bob McCabe learned of my difficulties and sent me a top-entrance trap. I climbed the measly little jack pine in which they were nesting, put the young in the trap, set it on the ground and caught both parents in a matter of moments. They *were* our shrikes.

One of the adults hardly looked like a shrike. It was deformed. Its bill was crossed, but the bird was fine and fat and went right back to feeding its young after I returned them to their nest.

Later that same summer a lady was poking about in some trash barrels back of a filling station in Princeton, Wisconsin. I have no idea *why* she was doing this, but what she found was a dead bird wearing one of my bands. She examined a number of bird books and field guides, but was unable to identify the "crossbill." So she sent an accurate and lengthy description of her find to the Feds.

Ah ha!

DEAR MRS. HAMERSTROM,
A bird, wearing one of the bands issued to you, does not answer the description of any native wild species. We presume you have banded an exotic cage bird.

GENTLEMEN:
My band was placed not on an exotic cage bird, but on a migrant shrike—the only shrike with a crossed bill that I have encountered. It was simply a monstrosity.

My answer ended the case of the cross-billed shrike.

I do make mistakes in my reports. It is human to err. I console myself: banks make mistakes too. That rough-legged hawk with the red bracelet and the shrike with the zany bill were *not* mistakes, just persecution by Fate.

Monica's come-on

Birds die in a variety of picturesque manners. Kestrels and barn swallows often enter abandoned houses through a broken window, flutter about trying to find their way out and eventually a skeleton, perhaps bearing a band, is found in an upstairs bedroom on the floor among old newspapers and discarded shoes. Goshawks are apt to crash into or *through* windows, suffering brain concussion or instant death. (I think they do this because they are attacking their own reflections, but I could be wrong.) Just at the time that DDT spray planes were causing breeding failures in birds of prey, a harrier flew into a spray plane, like a kamikaze pilot. I learned this by asking questions.

The bander, particularly if he is a specialist, wants to learn how long the bird has been dead. A bird found *fresh* dead—and with good evidence, such as "maggots in carcass" or "shot on opening day of duck season"—can yield excellent records on longevity.

Most banders grumble about the public and complain that they have great trouble getting finders of banded birds to answer letters about just where and how the banded bird was found.

I have no trouble.

I use every scrap of information sent me by the Banding Lab. If the report was originally sent in by an official, I use my most official-looking stationery and type a straight business letter. But if the band was turned in by a member of the public, I get out a cheap yellow lined paper and write a long folksy letter in pencil. The letters vary, but run something like this:

DEAR MR. SMITH,
 I'm so glad you got my banded bird. It was in a nest in a tall oak tree (quite a climb). I found the nest in deer season (it is close to a good runway), and when I went back to check it out in spring—sure enough—there were the hawks.
 What kind of gun do you shoot? I have a double barrel 20 for birds and a 250:3000 for deer.

41

Please tell me more about how you got the bird. All the way to Alabama is a long way. If you can give me the date you got it, we can tell how quickly it got all the way to you.

If you found it dead, had it been dead long? Was it still stinky? Or all dried out?

I banded three hawks in that nest, but yours is the only band I've heard from. I am looking forward to hearing from you. Yours truly,

And I look forward to seeing my self-addressed envelopes arrive, usually containing just the information I want. As a sample:

DEAR SIR:
I was in the woods hunting on January 1 and saw the hawk flying overheard, so I shot him and when I picked him up I saw the band on his foot. He was killed near Kingville. He was flying North East. What business is it of yours?
Sincerely,
BOBBY SMITH

Sometimes I just wait until the rates are down and telephone. My most recent exciting recovery came from Cuba! A harrier, banded as a nestling, was reported taken in Cuba in the fall of 1980. I studied the best map I could find and discovered this amounted to an over-water flight of some ninety miles! It was my farthest record for this species. Impatiently I waited for five o'clock, alerted an international operator, and struggled to put a call through to Señor G. in Sancti Spiritus. I tried over and over again, having decided to spend almost five dollars to verify that band number or get parts of the bird if possible. No one told me—least of all the many telephone operators—that the political situation at the moment in Cuba was tense. Desperate relatives were trying to reach each other, various characters of the wrong sort were undoubtedly trying to do the same; telephone lines were jammed. My chances of confirming that band number that night were nil. I gave up after midnight.

That is, I gave up telephoning. A letter would have to be written. I have perfect confidence in my ability to talk with someone in my peculiar, vivid, ungrammatical version of what I call Spanish. My Spanish produces gasps, gales of laughter, polite suggestions on how to improve my language, but my Spanish is understood. The telephone held no fears for me. But I have no notion how to write a single sentence in Spanish—much less how to compose the kind of letter on lined yellow paper that will be answered.

Something would turn up. And sure enough, very soon Monica Herzig, a brilliant Mexican graduate student, arrived.

"Monica, you must help me."

"Yes, Fran."

"Write a letter to Señor G. It must be a good come-on. Do you know what a come-on is?"

Monica's inscrutable, beautiful face turned on its Mona Lisa expression.

I breathed deeper. "Tell him his data are very valuable to me."

The Señor did indeed answer. He had kept the band and verified the number for me. He had shot the bird while hunting, but saw no reason to keep any part of it because that kind of bird was not fit to eat. Thereafter the letter got very flowery. The Señor wanted something in return for those very valuable data. "My government will have no objection if you send me one." One what?

Finally we deciphered the rest of the letter. All he wanted was a reward for turning in the number. In fact all he wanted was a brand new shotgun.

Eagle in a peach tree

Frederick and I had made ourselves egregiously unpopular in 1959. We were standing up for a principle: namely that trees should *not* grow wherever a tree can be made to grow, but that Wisconsin's public lands should have the rich diversity of various stages of the plant succession, supporting many species of wildlife — including the sharp-tailed grouse.

Anyone who has been in public service knows what these trying periods are like. Friends take you aside and admonish you in low tones, the uninformed know there is something up and don't know whether or not to be seen speaking to you, and the opposition plainly desires to pelt you with rotten tomatoes — or worse.

Just at this time the State of Ohio borrowed us from our own conservation department, to make a survey of potential sharp-tailed grouse range in Ashtabula County. Ohio rolled out the red velvet carpet. Frederick was given a plane and a pilot; I was given a plane and a pilot to make it easy for us to map likely looking sharptail habitat from the air. We were housed in a plush motel. And, furthermore, Ohio biologists and some federal personnel were sent to our headquarters so they could have the valuable experience of meeting us. It was heady, especially in contrast to the rotten-tomato climate at home.

After I had completed my survey (which went smoothly because my pilot had not gotten lost), one of the biologists mentioned that there was an eagle's nest nearby.

"Eagle! We ought to band the young."

"Waal, who's gonna climb that tree?" he drawled in what I assume was Ashtabula County dialect.

"What kind of a tree is it?"

"It's a *peach* tree."

Knowing full well that there isn't a peach tree in the world that can't be mastered with a fourteen-foot stepladder, I spoke without hesitation. "I will."

The chap looked me over with considerable interest.

"I have bands. Frederick has to go up with his pilot again tomorrow, so I'm free. Shall we start about eight?"

He shook his head slowly, walked a few yards, and then turned around to look at me again. "Eight o'clock," I shouted after him.

He nodded.

He was ready the next morning. Just in case there was more than one eagle's nest, I stowed my climbing gear in his station wagon, which was so full of biologists that I had to ask one of them to help me find room.

We drove through pleasant Ohio woods, crossed the line into Pennsylvania and parked near a dock, where I recognized quite a few of the federal personnel on hand. Everybody piled into a large motor launch. It did not occur to me to wonder where they were going or why. . . .

I watched the shoreline for peach trees. Pymatuning Wildlife Refuge abounded in tall hardwoods and conifers. Finally I asked my guide, "Where is this peach tree?"

"On an island. We'll be there pretty soon."

So I started watching the islands. When the launch was beached on an island everyone disembarked. Before I had time to ask, my guide pointed. "There it is."

High in a sky-busting *beech* tree was an obviously active bald eagle nest.

I have good professional climbing equipment: climbing irons with long spurs, a tree surgeon's belt, and a long sling to go around the tree.

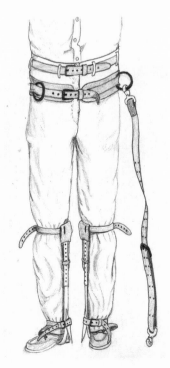

My sling was obviously far too short to go around that tree trunk; besides, the giant beech was split about two-thirds of the way up, and I had no way of roping in to get past the split!

A tall nearby hemlock was my only chance of getting anywhere near the eagle's nest. I fastened my banding kit to my belt and scurried up the hemlock to reconnoiter. There was a possible route to the crown of the beech. I didn't like it, but it was possible.

One beech branch could be reached by crawling out on a hemlock limb. I hitched myself out until my weight brought the hemlock branch down so I could reach the beech branch, which was bigger than my wrist. I knew it was strong enough. The question was: could I get back? Reassuring hemlock branches below me suggested that, with reasonable luck, the beech branch would bend so I could reach some hemlock branch to take me back to earth after the eagles were banded.

I would never have tried it if I had been alone, but it was plain that all those delightful men had come to watch me. The honor of the Hamerstroms was at stake. Besides it was long before the days of Women's Lib and not a time to chicken out.

The beech limb swooshed down with my weight, and I slithered over to the main part of the trunk like some kind of fairly awkward mechanical toy. It was no problem, and it was no problem to get to the eagle's nest. I grabbed for the bigger eaglet, but she backed off to the far side of the nest, so I banded the easy one. I had to take the nest apart a bit to find a stick with a hook at the end, with which to horse that big eaglet to where I could grab her. For a moment I looked down to see how far she would fall if something went wrong.

Far, far below me, the motor launch looked small. And on the beach there were faces—little faces of all those men watching me. I giggled.

Then I heaved myself onto the edge of the nest and pulled with my stick until that big baby eagle lost her balance. For a split second, her left wing tip was almost within reach. I lunged, grabbed it, and pulled her unceremoniously across the nest toward me. I rested for a moment or two after I finished banding her. Then I climbed slowly down to a certain slender beech limb—my lifeline. When I reached it, I rested again with arms and legs hanging down to bring a good blood supply into my muscles. I leaned against that beech limb and I didn't like it.

It does no good to reflect too long. I hitched my way back across my natural bridge with butterflies in my stomach until the slender beech limb bent. I moved farther, and it made contact with a perfect hemlock branch. I was home!

Then I rested again, not because my muscles were tired, but because I had to put distance between myself and recent terror.

46

When I got down, my guide said, "You'll tackle anything, won't you?"

My answer must have puzzled him. "You have unusual accents around here."

Some peach tree!

Gabboons

As I was approaching my stuffy forties—an age when many people tend to become unduly conservative—a disreputable black sedan pulled into our driveway on a bitter February morning. The fenders were dented, several windows were cracked, and one window had apparently been rolled down. (Later I learned that it had not been rolled down. It was just gone.)

A motley crew, in tattered field clothes and wearing home-made coonskin caps—excepting one, who wore a broad-brimmed velour trimmed with an ostrich feather—emerged from beneath quilts and got stiffly out of the car.

The leader said, "Say, can you come eagle trapping with us? There's lots at Petenwell."

My day was free. I liked the looks of that bunch. I had never trapped an eagle.

"I'm sorry," I said, "I can't go for forty minutes." It was the beginning of an undying friendship, and we acquired some new gabboons—a politer term would be *apprentices*.

"You," I pointed at one of them, "will find four traps inside the barn, near the door. Bring them up to the house."

I turned to another. "You will find a feral cat on the floor of that station wagon. Please go get it."

"Somebody's pet kitty?"

"No! Feral means gone wild—roving the country—and scrounging what it can find to eat."

"What's it doing in the station wagon?" The boy stared at me.

"I need it." It took me some time to get used to picking up cats on the highway, but they're dead—belong to no one—and they're good hawk food.

"I need them to feed all those hawks and that eagle that I'm nursing back to health."

The boy had unblinking eyes. And what long lashes! His stare was disconcerting.

"Do you mind picking that cat up?"

"No."

"Well *do* it then, and put it in my VW bus. We're going in that."

I wanted to watch his reaction to the cat, but I have found that once an order is given, it is best to turn away—clearly assuming that obedience will follow. So I turned to the leader. "Let's get the rest in. They can make sandwiches and coffee." (I never like to go anywhere without plenty of food.)

"Have you got any fish?"

Jack* moved his head slowly from side to side. "I sure wish we did. All we got is pigeons."

Gabboons! Expecting to trap bald eagles with pigeons. I tried to telephone one of the two grocery stores in town. "What kind of fish do you have?"

It seemed to be a poor connection—or it could have been that the woman at the other end of the line was having trouble with my Boston accent. Enunciating carefully, I raised my voice.

"Fish. I want fish and I need them right now."

The gentle voice at the other end of the line gave me a lengthy explanation. I couldn't catch most of it because the gabboons were waving knives and arguing about how thick to cut the bread. But I did catch the last few words.

"We don't have *any* fish."

Finally out of patience, I shouted, "Don't be silly. Of course you have fish! You have sardines . . . I've seen them right on the shelf! What other kinds?"

"Is this Mrs. Hamerstrom?"

"Yes."

"This is Violet Hakes."

"*What?*"

"You know, your neighbor, Mrs. Carl Hakes. I'd give you fish if we had any, but we really don't."

"I'm awfully sorry. I have to go eagle trapping and I thought I had the *store*."

Blaming this unfortunate incident on my chronic ineptitude with telephones, I jumped into the car to go and buy fish.

Mr. Potton, who ran the store, was well aware that the Hamer-

* Jack Oar, who some twenty-five years later illustrated this book.

stroms liked plenty of fish in their diet. He was not prepared for today's request:

"I want fish with scales on them."

"*Scales?*"

For a moment I almost weakened to explain that frozen cod fillets and little chunks of breaded fish then coming into fashion would not be attractive to eagles. I needed fish that looked like fish, but that would have led to a lot of conversation. I bought two pounds of smelt with the heads on, and then I selected some long, thin, almost gold-colored smoked fish.

"Beer," he said. "A friend of mine near Green Bay smokes these and they are just right with beer."

"Thank you."

Mr. Potton didn't know that I don't happen to like beer, but he *did* know full well that those fish were not going to be served with beer.

He ran his thumb lightly along the scales of one of the long goldeny fish and studied my face. I just said "Thank you" again, paid, ran out of the store, and hurried home.

The gabboons were making a lot of noise, but they gave an impression of a job well done.

"Let's go! Where are the sandwiches?"

I was unprepared for the absolutely innocent, stricken look that as many as five gabboons can turn on in a moment.

"We ate them."

It was Lesson Number One for me: always assume that any food is going to be eaten unless it's hidden. I prepared another lunch by putting all the cheese we had, all the crackers we had in the house, and a couple dozen apples in a large paper bag.

Jack sat in front with me; Frank, a tall seemingly languid boy, assumed a pose I'd normally associated with Cleopatra, and fell sound asleep. The rest argued. They argued about the fine points of some erudite matter such as the juvenal plumage of the Swainson's hawk. They quoted authorities I'd never heard of until I went to graduate school! "Ridgway don't have it right," one shouted, "I can tell you where it's located at. Look at page 227 in Bent. That guy's got it *right!*"

The argument veered to African eagles, and again they knew the world authorities, and quoted them in English so ungrammatical that I sometimes had trouble following what they were saying. I'd never met people like this. Specialized, rowdy scholars — united by their dislike of high school, by their flamboyantly uncouth manners, and above all by their passionate interest in the birds of prey.

When, at last, the arguments died down I raised a question that had been puzzling me considerably. "How did you happen to come for *me?*"

Tom, whose gift for flattery is unsurpassed, simply said, "Everybody knows about you, Fran."

I looked at Jack and he said, "Well, it was this way. . . . We knew there's eagles on the Wisconsin River, so we went to Petenwell to catch them."

"What did you want to catch them for?"

"To band them. We've got the bands—number nines for eagles."

Much mystified, I asked, "How'd you get a banding permit?"

"Wrote for bands. Then some guy in Washington phoned and asked questions."

From what I'd heard from the back of the car, I deduced that those questions had been answered as correctly as any professor of ornithology could answer them . . . not as elegantly, but as correctly. (It was far easier to become a bander in those days. In 1957 banders of unprotected hawks and owls were held in such low esteem that they were simply issued numbers, rather than permits.)

"We got to Petenwell late yesterday and tried to figure out how to catch them. Bal-chatris, like the ones you and Berger published on, seemed the best idea, but the stores were closed so we camped up on the bluff."

"Sleeping bags?" I inquired.

"No, we borrowed quilts and rolled up in them. And then in the morning our boots were so stiff that we couldn't get them on. We ran stocking-foot down to the car and hightailed it for Necedah to look for a hardware store. Found it."

There was the rich aroma of cigar smoke in my car. I glanced in the rearview mirror and noted that Tom, who looked about fourteen, had lighted the stub of a plump cigar.

"Our clothes was all wet when we got indoors, so we rigged a clothesline and hung them up to dry by the stove."

"*In* the store?" I asked.

"Yup. We used a clothesline, but we didn't cut it."

"The store's clothesline?"

Jack nodded. "And then we *bought* hardware cloth and lacing wire."

Jack paused to let this virtuous deed sink in. After a while he scratched his head and admitted that the man in the store probably didn't like that they were using his tools. "That Raymond," he jerked his head toward the small chap wearing the ostrich plume in his huge hat, "was running all over the place finding tinsnips and pliers and hammers — and we didn't have many clothes on, because most of them were drying."

I began to picture the scene in a local hardware store that I knew very well. And I knew Mr. Swasey, the owner, who was very careful of his stock and weighed out nails as though each one cost a dollar. I giggled.

Much encouraged by this small sound, Jack became expansive. "He didn't know how to get rid of us and more and more customers were coming in. We'd told him we were trapping eagles so he phoned the game warden. I think he had a word with him about us."

"Tom there, he kept leaving his ceegar on the counter while he was wiring the traps together and it would burn little grooves in the wood."

I could almost see this crew—all over the store borrowing tools, shouting and arguing, half clad—and I could almost smell the steaming wool of clothing drying by the stove.

"Well," Jack continued, "the warden couldn't throw us out. There were too many of us."

Jack sighed, and I had the feeling that a good rough-and-tumble fight in that hardware store would have added zest to his day. "That warden was a *big* man." Jack sighed again. "He told us we couldn't band eagles because we're from Rockford, Illinois—not with nobody having a Wisconsin permit. We knew *you* had a Wisconsin permit." And then I got a sincere compliment: "We figured you'd come."

The Rockford Bunch is still a treasured, but sometimes demanding, part of my existence.

On the art of wearing out one's welcome

The bald eagles wintering below Petenwell Dam offered quiet frustration. They sat about in trees hour upon hour on end — looking statuesque. And when it suited the pleasure of an eagle she left her perch, circled over the turbulent water below the dam and scooped up a dead or dying fish that had been pummeled through the turbines. Her first wingbeat after the catch tended to be a bit clumsy, but in no time she had the fish held streamlined with her body and flew majestically to one of the many favored perches where she could take her pleasure on raw fish.

Raw fish are an excellent and nutritious diet for bald eagles. Our problem, however, was to offer something so enticing that an eagle would come to a given point — to a trap. We were understocked with first-class eagle bait on our first foray, so we garnished our smoked fish with nooses and put them on the beach. We made one very careful set with the largest smoked fish: its rear end floated in the

water and, after we had broken its backbone in several places, it looked severely crippled, but still alive. Snow spat out of a lead blue sky and our fingers were numb.

Two live pigeons were placed in a large bal-chatri and given a handful of corn to keep them happy the rest of the day.

The road-killed cat was frozen stiff, but Tom whacked it with a machete until some of its pale frozen flesh peeped forth beneath the fur. Then he wrapped the whole carcass in a poultry wire noose carpet after which he took off his mittens and delicately, with meticulous care, arranged every noose so it was upright and in perfect position to close upon the toe of any eagle that came there to feast.

We were short of traps and of cord to tie each set to a tree, but had extra smoked fish, so I suggested chumming. Strictly speaking, chumming is dumping a lot of entrails and other fish parts into the water to attract fish. But I wanted to make a dump heap of fish on the beach to get the *eagles* to come to one good place to feed. Jack and Elva made the fish dump on a sand drift. Coppery smoked fish made up most of the pile. It was time to move. I shouted, "Let's get in the car. We're keeping the eagles away!"

Tom was kneeling in front of his cat rocking to and fro. His mittens lay on the ground beside him, and his hands were out of sight against his stomach. Any woodsman knows that he was holding them close to his skin, waiting for the pain of near-frostbite to seep away from under his fingernails. Frank viewed the set with distaste. "The meat don't show too good."

Jack destroyed the cat set with one swift flick and grabbed the cat. Tom yelled, "Why You!" and brandished the machete.

Jack did a little dance on the sand and simpered, "We just need to get that cat meat showing."

Tom lunged. Jack side-stepped and ran, bouncing like a bronco.

Elva and I, unaccustomed to such people, watched the machete whizz through the air—just missing Jack's shoulder. The fight, now joined by the others, raged along the water's edge. Frank, now no longer sleepy, grabbed the machete and gave swift jabs like a fencer to keep the others at bay.

The fight subsided as swiftly as it had started. Jack carried the cat over to a log and tried to break it in half by kneeling on its head and its rump. Failing in this attempt he whanged it over the log. At last he gave up and fixed the noose carpet over the carcass with the same care that Tom had used over half an hour ago. I helped straighten out the nooses and soon Elva joined us. She murmured, "Do you see those men up there in that powerhouse? I think they've been watching us."

It was too far for me to see much of anything, but as we approached the car—to hide so the eagles could come back, and to get warm—I could see men walking about slowly and comfortably in shirt sleeves!

We climbed into the bus. As the heater wasn't working well, I waited patiently for our seven bodies to bring the temperature up to where we'd quit shivering and begin to breathe normally again. The windows steamed up. Elva kept a tiny area of the window, facing the river and our sets, scraped clear, so she could watch. Raymond opened a window facing the woods as wide as he could declaring, "I want fresh air." Tom banged it shut and a new fight started up in the back of the bus. I reached for the machete, laid it on the floor up front and firmly planted both feet on it.

It is curious how quickly one adapts. The fight continued to rage behind me and I sat in the driver's seat daydreaming. If only Elva and I were here alone, I could introduce myself as a biologist from the Conservation Department, and we would probably be able to get permission to watch our sets from the warmth and commanding view of the powerhouse.

Frank's drawl cut through the thumpings and grunts from the back of the bus. "We'd have a prettier view from up there."

Just what I was thinking. I offered to go alone to the powerhouse to try to make arrangements. Nobody heard me. They flung themselves out of the bus and rushed toward the shiny glass door entrance. By the time I caught up to them they were signing the guest register. I glanced at the names. Raymond had signed himself simply as *The Merlin.*

Above this someone had declared himself *Frederick the Second of Hohenstaufen,* then came *Simon Latham, Henry the Eighth, George Turberville,* and *Pope "Pious."* All the addresses were fanciful and spurious except for the last: *The Vatican, Rome.*

I had planned to sign the guest register as *Dr. Frances Hamerstrom* to give the eagle trapping undertaking a touch of class. For a fleeting moment I almost signed as *Diana, Goddess of the Hunt* (not being able to think of any famous female falconer). Instead I put the pen firmly back in its holder and refrained from giving out my name. I followed the trail of muddy water up the stairs to the office.

Frank was sprawled behind a shiny desk lighting up a cigar. Elva squatted on the floor and the men in charge of the powerhouse were answering her questions about eagles. Frank can make himself at home anywhere so he had already learned that in relatively warm weather, when much of the ice breaks up, the eagles disperse downstream. They congregate near the dam only when the weather has been well below zero for a number of days.

Two official-looking men sat behind desks and a workman appeared with a mop to clean up the muddy floor—our mud.

An eagle flew slowly over the roiled water and lazily returned to her perch. Planters of African violets along the office window contrasted sharply with the spitting snow beyond.

Tom prepared to flick a cigarette butt on the floor. I gave him a look, smiling sweetly. He ground the butt down among the luxuriant roots of an African violet in full bloom.

Petenwell powerhouse was to be our second home for *almost* all the remaining weekends of that winter.

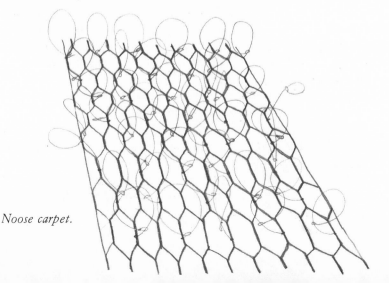

Noose carpet.

Back soon after dark

At noon, the men from the powerhouse produced neat black lunch boxes, laid paper napkins on their desks and started to eat. Our crew produced one paper bag stuffed with smashed crackers and a chunk of aged cheddar that I figured would last us a week. There was one bottle of pop. Jack and I barely managed to stave off another fight. The fight for a sip of pop.

Somnolence. Throughout the long hours of the afternoon, the eagles dozed and our semicomatose state was interrupted now and then by Willy Hillyer, who had worked at the dam for years; he alerted us whenever an eagle was airborne and within fifty yards of our water sets. No eagle showed an interest in our beach sets. It seemed they hardly ever ate, and only from the surface of the water.

At dusk, we went to pick up our bait. I looked at those sandy coppery smoked fish on the chumming pile, kicked them free and tucked them under my arm. They might come in handy next weekend. Next weekend? We all knew we'd be coming back next weekend, but I didn't know then that the fish were not destined to end up as eagle bait. It was good to be starting home. The Rockford Bunch had 210 miles to travel that night, and it was cold. The snow had stopped and stars promised a clear night. Elva and I had only forty-nine miles to get back to Plainfield. I thought of the note I'd left Frederick: "Gone to Petenwell. Back soon after dark."

It was soon after dark now.

Revived by the cold night air on the way to the car, the Rockford Bunch ran to where we had parked it. I followed more sedately, savoring the quiet—away from the steady hum of the turbines and the disorder we had created in the office of the powerhouse.

"We know where the eagles roost! Let's go!"

I almost had sense enough to suggest going straight home—but there still might be a chance of catching a *roosting* eagle. "We saw them last night when we were sleeping up on the rock."

For a moment I looked up at the black outline of Petenwell Rock silhouetted against the sky—at least twenty below last night and these boys had slept up there wrapped in quilts!

Common sense lost out. In a matter of moments, I had parked

the car near the base of the rock and found myself plunging through thin tinkle ice and clutching thorny catbrier vines to steady myself. I felt in my pocket for my compass, but at the rate everybody was going I had a simple choice to make: keep up with them or pause to get my compass direction. I chose to keep up.

Suddenly we were near eagles. Raymond fell down with a crash. Jack yelled, "Quiet!" and at least five eagles ponderously left their high perches and disappeared. This precipitated a conference. Everybody sat down. "There is a story," somebody said, "about two eagles that came right up to a camp fire and got caught."

"We could try it."

Cold worked its way into my bones. I got up to beat my arms against my chest and announced, "We are going home."

I had no idea which direction the car was, but Frank ambled away into the night with confidence and the rest of us followed him. In a mercifully short time we were back at the car. The conference continued.

"Fran, have you ever heard of candling?" Jack asked. "It's for getting back lost hawks. Somebody holds a candle on top of a long pole in front of the hawk's face and then somebody else eases a T-perch up under her breast so she'll step up on it."

I gave Jack a sharp look, but he wasn't putting me on.

He continued, "Then you move the perch down real slow and grab the hawk."

"It might work for eagles," I said, "but I think I have a better idea. Have you ever heard of the Humane Cat and Monkey Catcher?"

"Huh?"

I nodded. "Fire departments use them," but I refused to explain further.

Now that seven bodies were warming up the inside of the car, the smell of fish was becoming astonishingly pronounced.

"Did somebody leave those chumming fish near the heater?"

"Have some!" Raymond said, "Here's one just peeled."

Gratefully I accepted a slab of cold, not very sandy smoked fish. When we got to our house, I was far too tired to notice that the back of Frederick's vehicle was liberally strewn with sandy fish bones, golden scales, and squashed heads.

Frederick was already in bed. With much slamming of doors and shouting, the Rockford Bunch departed, yelling, "See you next weekend. We'll be up Friday night."

It was 1:00 A.M.

Please collect burned-out light bulbs

The idea of catching eagles at night had a strong appeal for me. At my earliest convenience the next day I dug out a tree surgeon's catalogue, looked in the index, and found the item I wanted: the HUMANE CAT AND MONKEY CATCHER.

I very much doubted that any wild eagle would let itself be lowered to the ground, with nothing to hold it. We needed a device to grab it, and the catcher had all the requirements. It seized at the pull of a lever and had an adjustable stop so that whatever creature was being hauled down out of a tree at the end of a pole would not be grasped too tightly. It was used to rescue pets that had climbed high and either didn't dare come down or preferred not to. We could add a weight to the long pole, grasp the startled eagle with this fine contraption and easily run it down.

Then I looked at the price. Twenty-five dollars. Heavens! I had bought a good second-hand car for twenty-five dollars. The price was out of sight.

Reluctantly I set aside all dreams of catching at night. It was time to call in more experts and devise a way of catching eagles fishing on the surface of the water. I dashed off a note to Berger and Mueller:

> Thirty eagles (20 adult and 10 immys) are concentrated at Petenwell Dam. Bring traps and bait. We have a blind. It is 288 feet long, with comfortable chairs, and decorated with blooming African violets. Let's meet Sat as early as possible.

All four Hamerstroms set out well before daybreak on Saturday. We arranged some beach sets, and settled into the powerhouse. Fred-

60

erick's presence must have been a relief to the personnel there. He introduced himself properly, and asked permission to watch eagles from their establishment. Alan had brought some books to read and Elva brought crayons and drawing paper. Frederick and I watched the eagles and all was quiet.

Helmut and Nancy Mueller and Dan Berger arrived well after sunup. Their arrival created only a small stir and was appropriately sedate. They learned what Frank had learned: the eagles concentrate only when the weather has been really cold—fish only on the water—and last winter they had fed on a dead deer body.

Helmut mumbled, "We should try that."

Just then there was a disturbance down by the entrance. Somebody yelled "Odin!" and there was considerable scuffling.

Elva ran down to see what the commotion was all about. The racket moved on up into the office. Two of the Rockford Bunch carried half a dead deer dripping blood across the immaculate floor, and two more carried a struggling Elva, shouting, "*This* would make good eagle bait too!"

. Jack added, "A pity she isn't a little baby."

Willy showed us where we could scoop up some live fish for bait, but they were small. Helmut, Dan, and Frederick studied the ways of the eagles, and by the next Sunday our pattern for winter weekends was established: we arrived before daybreak on Saturday and on Sunday to put out ever more complicated sets.

I telephoned the Rough Fish Camp of the Conservation Department and asked for some big fish. They said a truck was coming over our way anyway, and delivered some 200 pounds of huge carp frozen into one solid mass.

We pried them apart and stuffed some with burned-out electric light bulbs to make them float. Then they were carefully wired so nooses could be firmly attached. We brought our rowboat to the waters of Petenwell so we could get our floating fish sets out onto the river.

Berger left the boat at the water's edge without tossing the anchor ashore and the next time we looked it was gone! Apparently more water had been needed to accommodate the turbines so our boat just floated away down the Wisconsin River.

Frederick called the game warden for help. The warden miraculously managed to find our precious boat and, after he had beached it, he came up to the office. The scene that met his eyes was not unfamiliar. The Rockford Bunch lay about on the floor smoking and eating peanut butter and jelly sandwiches, and a pair of socks was drying on the back of one of the office chairs. It wasn't quite as messy as the scene in the Necedah hardware store, but the same people were involved.

The warden addressed the group at large. "I'd like to see your banding permit."

I produced my permit and thanked him for saving our boat. He looked pretty uneasy, but finally left saying, "I guess it's all right."

Our sets were a marvel. Each fish was stuffed with a light bulb to make it float and a weight in the belly area to make it float upright. Furthermore, each had been tested in a bath tub and floated nicely in still water, but the current of the Wisconsin River sucked them under when we tried to anchor them. First we planned to build small rafts to keep our fish above the surface, but we suspected that they would look unnatural and keep the eagles away. Then somebody had a better idea. We anchored duck decoys instead. Duck decoys floated innocently — among other ducks — on the turbulent waters; each decoy so to speak towed one long cord to which six or eight floating fish, adorned with nooses in catching position, were firmly fastened.

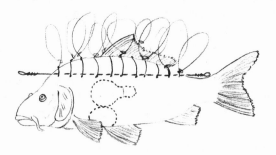

Deer hair has an annoying way of sticking up unnaturally if one tries to hide anything in it. With ever-increasing desperation we tried to hide two noose carpets in the stiff, stubborn hair of the front half of the dead deer on the beach. Finally we settled on tying the two noose carpets to her neck and lightly burying them in the sand, so that any eagle that walked around the carcass had a fairly decent chance of getting caught.

Helmut and Frederick called a conference to establish rules for future eagle trapping. By now about twelve trappers convened every weekend.

Helmut announced, "I don't want any monkeying around out-doors. Everybody either stays inside or goes to the parking lot and uses the regular road to get out.

"Our first responsibility is to. see that no eagle gets drowned." Frederick suggested posting one observer at the top of the high stairs at the west end of the huge building. A spidery iron staircase led up to a small platform where one person could watch all the water sets. From this vantage point the official observer was to give the alarm whenever an eagle struck one of our floating fish, and two others were selected to rush out and man the boat to rescue any captured bird in short order. The change of guard was established for twenty-minute intervals. At first, everybody watched and gasped each time an eagle left its perch to fish, or simply to shift position. It seemed strange to watch the workmen wander about their own business, oiling machinery or checking dials, but as the hours went by, each twenty-minute shift seemed like a long time.

The conference resulted in another ultimatum: all sets are to be put out before daybreak.

The men at the powerhouse had no trouble hearing everything that the Rockford Bunch said. Even their whispers could be heard for yards. Frederick does not raise his voice, but speaks with quiet authority. After the gabboons shouted their farewells and romped down the stairs, Frederick thanked the powerhouse crew for their courtesy. They must have thought him reassuringly normal, at least until he turned to Helmut and said, "Please collect burned-out light bulbs."

Barack.

Just one apple core

Berger and the Muellers manned the rowboat well before day-break. The temperature was probably about thirty below, but they put out all the water sets and arranged every noose in catching position with bare hands; wearing mittens, there is no way to make a nylon noose stand up properly. Naked fingers are the only solution. All the gear was wet, and when they came into the powerhouse they kicked at the door until one of the workmen heard them and opened it for them. Their hands were stiff-fingered and incapable of opening doors. Their faces were set—fighting back the pain of frostbite. Frederick asked, "How was it?"

Helmut's answer was, "Bad."

When Helmut admits to any pain it must be very bad. The fingers of the water-set contingent swelled, turned rosy, and gradually subsided. The men at the powerhouse offered various home remedies; all were stoically scorned. Finally one said, "You care a lot about catching those eagles, for sure," and I realized our group was gaining their respect.

We persisted. And as we waited in vain for a shout from the high platform, boredom set in and we tried to devise new methods of catching these seemingly phlegmatic birds.

64

The Rockford Bunch wanted to try a "barak," so they brought along a trained red-tailed hawk and tied a moderately heavy ball adorned with feathers and nooses to its feet. They proposed to fly the redtail over the water in such a way that her seeming booty would be irresistible to the eagles. These would try to rob the redtail and tumble downward with it and get caught. Downward would be into the turbulent water, so we vetoed this method as too dangerous to both birds.

No one vetoed my suggestion. I tethered a moderately tame golden eagle on a sandbar well below the dam and placed a dho-gaza safely above her head. Two of us stood shivering on a bridge and watched, hoping that one of the bald eagles would dive bomb her and get caught in the net.

Golden eagle and dho-gaza.

We had had ample opportunity to learn that very little traffic passed over that bridge, but about an hour after we had everything in position, a whole caravan of cars parked near the bridge and masses of people started sauntering across, pausing to count and admire the eagles. Scopes were set up and binoculars came into play.

It was the annual eagle-watching-weekend of the Wisconsin Society for Ornithology!

As the bridge was covered with miscellaneous people anyway, I joined the group in time to overhear a lively argument.

One small boy exclaimed, "There's a golden eagle down there!"

His mother pulled him along by the wrist. "No, dear, it's a bald eagle—a young bald eagle."

"It's a *golden* eagle! I know!"

Not wishing to interfere, I followed them, miserably meek.

At last she let go of his wrist and I beckoned the child over to the guard rail. "That one?" I asked, pointing to our set.

"Yup."

"You're right. It's a golden eagle. It's mine."

Ed Peartree, the leader of the ornithology group, was watching me with interest. He had considerately *not* pointed out our set, knowing full well that his group would refuse to budge from any eagle-trapping venture. It was with considerable relief that I watched the cavalcade of cars disappear. We gave my eagle a full day to attract other eagles, but even when I gave her some meat to feast on in the presence of the wild eagles, she was ignored. It was a good try anyway.

Boredom created other problems. Not only were there more innovations, but there were also more fights.

We took stock of our accomplishments. And then I decided to try a new device. There was one favored eagle perch low enough so that I could reach its top. It was simply an old stump sticking up out

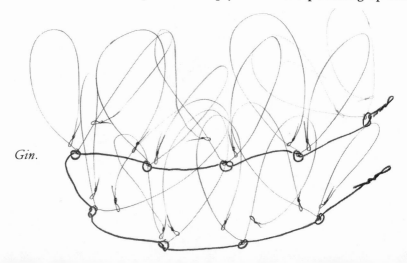

Gin.

66

of the nearby marshland, and I figured that a gin might work.

The gin, as far as I know, is best known to the poachers of Scotland. It is a set of nooses strung along a cord. It tucks nicely into a pocket and its uses are flexible. Ptarmigan, out for a morning stroll, can get their little feathered feet caught in it. The unwary rabbit, hopping down a bunny trail, may come to a swift stop.

My gin undoubtedly differed in a few respects from those abhorred by gamekeepers, especially as I was using modern materials. I took a two-foot length of stovepipe wire (which is almost as flexible as stout cord) and tied a half hitch about every three inches. To each half hitch knot I fastened a pair of fifty-pound test nylon nooses. Then I fastened the finished gin securely to a cord that ran down to the ground and was firmly tied there, and draped the gin around the top of the stump so that any eagle that cared to perch there might well get a toe caught.

I set that gin over and over again. It became a sort of ritual.

One weekend we picked up later than usual and my gin was gone! As the gin had never caught anything, my first thought was that somebody must have stolen it. It was too dark to see much of anything, but I tromped around in the marsh grass trying to find the end of the gin cord. My leg bumped into something large, soft, and alive. There was a snapping and clattering and I looked down into the eyes of a big female great horned owl!

I disentangled her, untied the gin and put it in my pocket, and using the madonna position I carried the owl back to the powerhouse to measure, weigh, and band her. (The madonna position is simply holding a large bird of prey cradled in one arm. The feet are firmly clutched in the hand, and it pays to tilt the bird's head away from one's body in order not to be bitten in the bosom. It looks hard, but actually, once mastered, it is easy, and with a little practice one can drive a car while madonnaing an eagle.)

The workers at the powerhouse had watched our efforts for so many weekends that they must have wondered if we were ever going to catch anything. Seeing a female member of our group nonchalantly carrying a huge owl like a baby undoubtedly restored some of their faith. It was further restored when Frederick produced gram scales, millimeter rulers, and an official-looking notebook.

It wasn't until their attitude *changed* that I recognized what their attitude toward us had become. The eagle trappers had lost their novelty and had become an ever more annoying nuisance.

The foreman took Frederick, Helmut, and me aside and announced, "Tomorrow there will be an inspection. Will you," he gave a little cough, "be coming here tomorrow?"

I suspected that he actually had been about to ask us *not* to come. He sighed and added, "The inspectors will arrive about ten o'clock in the morning. We have everything cleaned up."

I gathered the Rockford Bunch together and gave them the word. We do not have a *right* to be here, but you all behave as though you owned the whole powerhouse. You mess up the floors with food and wet clothes, and you leave your cigar butts under the desks, on the tables, and in the African violets.

Jack interrupted quickly, "We can stuff them into coke bottles tomorrow."

"No," I shouted, "that doesn't look nice either. No bottles and no eating until after the inspectors have gone.

"Do you realize they had to clean the floor three times after you all made such messes today? We are wearing out our welcome. Everybody's tired of all the mess and the swearing and the fighting. Behave!"

(Our son Alan had basketball practice and wasn't present, but I wouldn't have bothered to give him the word if he had been. He was by far the quietest and best-behaved member of the whole undertaking.)

The next day: inspection. Three men in business suits came up the stairs. No food was in sight, no wet clothing, and Tom dexterously pitched an almost completely extinguished cigar butt into a wastepaper basket . . . where it smoldered. . . .

Before the inspectors managed to mount the flight of stairs to the office, I removed the two most picturesque hats: Jack's homemade coonskin with a nasty-looking untanned skin inside, and Raymond's huge cavalier's hat with the ostrich plume. I hid them in the ladies' room.

One look at the august inspectors gave me the uncomfortable feeling that this was no normal inspection. Somebody had squealed on us, and we all knew who it was. One—just one—of the workmen at the powerhouse didn't like us. We called him S.P. to his face, and never told him the S.P. stood for Sour Puss.

The inspectors moseyed around and I was proud of the whole bunch. We talked very scientifically whenever they were within earshot. The wastebasket with the cigar butt ceased smoldering and finally, to our great relief, the three men went to observe the other parts of their domain.

This occurred just as Alan was up on the high balcony watching. Alan (and nobody had told him not to eat) had laid an apple core on the railing. He bent over to get a better look at the inspectors. The core lost its balance, gyrated downward and hit one of the inspectors on the shoulder. S.P. went into earnest conversation with the three men in business suits. They talked and looked up at Alan. Then they talked some more and looked at him again.

When the inspectors finally left, we broke out the food and settled on the floor to relax.

Then Willy came to give us the word.

"We are very sorry. You can never come here again. If that boy only hadn't thrown that apple core. . . ."

Frederick admitted he could use his influence to try to get us back. "But is it worth it?" he asked.

"We've had over a thousand nooses set for those eagles. We've had four mediocre strikes: one on a beach set, two at floating fish, and the only good strike was at a duck decoy without nooses.

"Let's do something else."

Neither he nor I realized that the something else would also involve the Rockford Bunch.

Have you seen a little girl?

Little by little I learned more about the Rockford Bunch. Their lives were dominated by birds of prey—at the expense of schoolwork, normal good manners, and creature comforts. And they almost always forgot to bring any food along. Even when they went to trap at Cedar Grove Ornithological Station, they failed to bring food. Berger and Mueller simply suggested that they camp atop a bluff by Lake Michigan for they had not yet earned the right to trap at the station.

Camp on top of the bluff they did . . . in an abandoned cemetery . . . which abounded in wild asparagus. They picked asparagus and, because water from the lake was a long way down, they cooked it in beer—with which they had not failed to provide themselves.

Not long after this episode, when I was trapping at Cedar Grove, the Rockford Bunch said they needed me to go bait-catching.

"Why me?"

"You can get us into the barns. We need somebody like you."

I looked them over and could see the farmers' viewpoint. If I were a farmer, would I let a group like that into *my* barn?

"The Hamerstroms trap their bait birds. Why don't you?"

I had expected the answer to be that they had forgotten or some other weak excuse. Tom answered, "Trapping is too slow."

Too slow! If this bunch had discovered a method of getting bait more quickly than trapping, nothing would deter me from learning it.

"I'll come."

These natural con artists piled into my car. Frank usurped the driver's seat and off we went into the night at a rate that I found alarming, until I noticed that he was an excellent driver.

Soon we drove into a farmyard. "You ask."

I knocked on the door and turned to look at the car while I was waiting. There appeared to be *one* person in the car; the other five were out of sight!

Plainly it was easier to get permission for two people to get into that barn—and one of them a female.

"Good evening. You have starlings roosting in your barn. We'd like to catch them."

The farmer, thin-necked and bald, scratched one of his very large ears. "Don't know as I want anybody around in my barn at night."

"Starlings," I added firmly, "mess up the hay and they have been known to carry hog cholera."

"Huh?"

"We would be very glad to take the starlings away for you."

The farmer said something I couldn't understand at all—a sound rather like um—ka—huh.

I didn't dare ask what he said, so I said, "Thank you very much."

Uneasily I waited for his answer. It came clear and distinct. "I hope you get them all."

As soon as I got down the porch steps, he turned out the yard light. I didn't think it very polite of him.

At that time I hadn't learned that yard lights (and full moons) are a menace, enabling the birds to find the windows and fly out before they can be caught.

Joyously, in the dark, I fumbled my way back to the car. Jack gave orders for my benefit. "Everybody is to climb as high and fast as he can. Nobody uses a light till I use mine."

The architecture of Wisconsin barns varies only slightly, but it is best studied in daylight. I followed the sound of pummeling feet past a long line of cows, scrambled up a ladder to the loft and then followed the clink of somebody's boots on the loft floor until I could grab another ladder. Swiftly I swung up the ladder for about twelve feet where I bumped my head. Two planks were in my way, but a rope was handy to get past them and continue up the ladder.

Feeling that I had lost time by stupidly bumping my head, I climbed full steam until I received a kick on the jaw from a cowboy boot. Much shaken, I stayed right where I was until I could see Tom slide a leg over a high window sill to get into catching position. I made for the other window and flashlights seemed to start blinking all over the top of the barn. Starlings, pigeons, and sparrows flew to the lights. Tom had arranged his light so that he had both hands free and was grabbing birds out of the air and stuffing them into a bag at his belt.

Nobody had thought to give me a bag, so I took off my socks and stuffed starlings into them and simply deposited pigeons in my pockets and inside my shirt.

"Who'll shake the track?"

Raymond yelled, "I will!" He scrambled clear to the peak of the loft and shook the long iron bar that acts as a track for the hay loader. Soon even more starlings, startled by the racket and their jiggly substrate, flew toward the lights.

Suddenly it was all over. Somebody shouted, "Let's leave the rest for seed." I was the last one down out of the barn and when I got back to the car everyone was sitting very quietly. We drove out of the driveway and parked on a hilltop to count our catch.

Twenty-three starlings, five pigeons, and four sparrows. "Pretty good."

"Fran, how did you get us into that barn? He *always* says no."

"Hog cholera."

"What?"

Using the silly simper that Jack uses when he is being especially annoying, I asked, "Don't starlings carry hog cholera on their feet?"

These people never applauded. Approval was demonstrated by loud roars accompanied by beating on the chest.

Much relieved that the foray had been a success, I longed to get back and crawl into bed where Frederick was waiting for me. But a conference was underway.

"We have almost enough."

I wondered how far it was to warm blankets and sleep. This was but a vain dream.

"We need to take just one more barn."

To my relief, Raymond said, "We'll never get permission if we wake anybody up. The farmers have all gone to bed."

But Frank had another idea. "I know a barn that's sort of far from the house. They'd never know we was there."

It was my car, but I wasn't consulted. We went haring off across the countryside, parked the car on a main highway, and traipsed up a bank in the darkness to a huge and ancient barn.

"No lights!" came an abrupt command. I flicked off my flashlight and stumbled along with the others in the darkness.

The moon appeared for a moment and the gaunt old barn gleamed silver. I found it beautiful but Frank sputtered, "Damn!"

Moonlight could undo our undertaking.

This barn was easier for me because I knew what to expect now. The crew scrambled high onto small platforms by the windows. After I had worked my way to a high gap where night sky gave a faint light thanks to a missing board, I could hear traffic far below. It hummed on the highway.

Then lights seemed to shine from every corner of the rafters, and, bird after bird, we made our catch.

Just as I expected somebody to yell, "Time to shake the track?" Jack called "Lights *out!*"

The outside of the whole barn appeared to be surrounded by waving lights. Raymond, who was near me, hanging by one hand and reaching for some sort of a foothold whispered, "The cops!"

Trembling, not daring to shift position, I stayed motionless with a good handhold just out of reach. I could see police, with gleaming badges, shining their lights around the barn floor.

Jack stood alone near a decrepit wheelbarrow that someone had left near the big barn door. Strong police flashlights seemed to pin him to the spot. Then he spoke in a slightly worried tone.

"Have you seen a little girl?"

"Is she lost? Can we help you?" the police inquired.

Jack is slightly wall-eyed and has an engaging innocent face.

Above him we clung to our catching perches without a sound.

Jack kicked at a small pile of hay and answered easily. "I guess she's gone home."

The police departed.

We climbed down as quietly as we could from our high perches, slipped away to the car and drove off without unseemly haste.

About two miles down the road we felt safe. The yelling and thumping on chests started once more, this time reaching a strange savage rhythm, interrupted by chants of

"Have you seen a little girl?"

When I finally crawled into the sleeping bag, Frederick turned on some light.

"Good Lord!" he asked, "What's happened to your jaw?"

"My jaw?"

He examined my face. "You're all black and blue!"

"Oh, that was a long time ago."

The police lights shining all over that barn, except up into the rafters, were still a vivid memory.

"Would you like me to tell you about a little girl?"

"Some other time."

Blow in their faces

"Is it dangerous?" People keep asking this question and often we are not quite sure what *it* is. Camping alone in far parts of the back country? Frederick and I feel safest far from cities and their riffraff. The back country has some bad eggs too, so just on principle we hide our camp sites: behind a hill, down in a ravine back of a pine thicket; and no matter how tired we are, we turn the car around so if unpleasantness should arise, we're ready to leave *fast* if hassled.

We try to avoid trouble and I have news for Eddie Bauer, L. L. Bean, etc. It is perfectly all right for them to build bright gaudy-colored tents for kids who *want* to be found if they get into trouble. But we need a tent of a soft natural color to *hide* our presence. Only in the outback of Australia did I feel safer in a bright blue "modern" tent.

But some people really mean: are the hawks dangerous?

Few of them bite, and being bitten by a hawk hurts less, and is less dangerous, than being bitten by a medium-sized rat. The feet are another matter. The talons of a fairly large hawk can go all the way into your flesh (unless they hit a bone). Novices often want to wear gloves; they are useless. Strong talons can go through gloves like butter, and gloves make you so clumsy that you are apt to get *footed*. Speed, and knowing which way a raptor can strike, gives you maximum safety.

A well-meaning member of the public rushing up to help you can cause a hawk to strike in an unexpected direction, and the next thing he knows, I am saying in a very even tone, "Please pull that talon out of the palm of my hand." I recall one chap who kept dancing around and wailing, "Shall I pull off its head?"

Banding headless hawks is not one of our specialties. But we have all gotten rather good at pulling talons out of each other. There are certain rules:

Keep away from the other foot (there is absolutely no point in having a hawk hold onto two people at once).

Get the hind talon out first and pull, not up, but out the way it went in.

Remember that 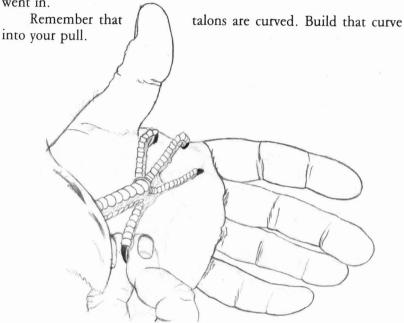 talons are curved. Build that curve into your pull.

As long as the hawk is still trying to foot you, each talon pulled out nestles firmly in the bottom of the hawk's foot and is not likely to do further damage. No raptor can strike with one talon at a time. But don't let down your guard when all the talons are out: a hawk can strike again.

Most puncture wounds don't really hurt much but, if a big hawk has struck a nerve and starts kneading, the pain is fierce. Our puncture wounds seem to heal nicely without infection. One gabboon worried me: even his scratches got infected. We keep up our tetanus boosters on principle.

It isn't always the public's fault that we get footed. And it isn't only getting footed near a nerve that can cause excitement. Once, it was time to do the laundry on a very hot day in South Texas. We pulled into a laundromat in Riviera, loaded the washers and congratulated ourselves: that sunny blazing heat was no good for trapping Harris' hawks, so we weren't wasting any time. After both machines finally stopped, we pushed those little wagons around and filled three dryers. Being rather bushed, I perched on a wobbly stool and contented myself with watching our clothes go round and round.

Suddenly Frederick announced, "It's clouding over!" A change in weather often brings the best trapping: hawks that have perched indolently, with one foot tucked under breast feathers, suddenly have a change of mood and are prepared to hunt.

I jumped off the stool. "Why don't you bait up? I'll cope with the wash."

Frederick moved in his normal, deceivingly deliberate but swift manner. By the time I unloaded the dryers and had stuffed everything back into the laundry bag, he had traps ready with pigeons and starlings. We headed south on Highway 77, where trapping used to be good before so much brush was cleared away. I sat, as usual, on the scale box in the back of the bus with the laundry bag stuffed out of the way. And he drove at the perfect speed for spotting hawks.

"Fran, those clothes can't have been dry?"

"They're not," I answered, "but everything dries in Texas." I pulled the laundry out of the bag and spread *his* all over the back of the bus to dry: shirts, socks, and immaculate underwear.

"Drop!" he called. And again, "Drop."

I hadn't seen the hawks, but after each command, a bal-chatri flew out of the door to repose by the roadside.

Instant reactors! We caught both before he had a chance to turn the bus around. I had no trouble getting them off the traps, no trouble weighing them in the tubes. It was when I was taking a routine wing measurement that one foot of a big female Harris' hawk slipped free and got me.

She nailed an artery.

Bright blood fountained in spurts, leaving tiny red droplets, which turned pink on Frederick's shirts, socks, and immaculate underwear.

The public quite rightly assumes that the moment of grabbing a hawk to get it off the trap has a certain element of danger. They are not tuned in to minor frustrations such as hurrying an untrained mouse to leave a trap.

This led to an unfortunate misunderstanding with the president of a bird club. He peered into the car and watched me using every trick I knew.

"How do you get them out of the traps?" he asked.

"Just blow in their faces and they'll go home."

What I didn't realize until much later was that the chap was referring to a male harrier that I happened to be holding like a bunch of flowers in my left hand.

Permanents

The gabboons found bait-catching such fun that they never devised a way of keeping starlings alive in captivity. After each trapping foray, they simply let their starlings go! (High-schoolers who are remiss in their studies have free time.) As Frederick and I were working for the Conservation Department and bringing up two children, I looked upon time as a precious commodity and soon learned to keep starlings alive.

Starlings like plenty of clean water or fluffy snow to bathe in. They will drink dirty water, but keep their feathers immaculate. They need dark cubbies to hide in, and only survive in well-lit cages if they are taken young, or gradually conditioned to spending time in broad daylight. They thrive on Gaines Homogenized Dog Meal, which has a high protein content, comes in small pellets, and is expensive.

Fresh-caught starlings do particularly well if they have a teacher. Any old starling who knows that Gaines Homogenized Dog Meal is good food sets the example, and the fresh-caught birds catch on.

Another trick to help starlings adapt to captivity is to put a pigeon in the pen with them. Pigeons are phlegmatic beings. Even fresh-caught feral pigeons adapt to captivity as though it were simply a more leisurely way of life. Now and again a pigeon is aggressive rather than phlegmatic. (Such individuals are tough, have a fine flavor, and end up in a stew.)

During the early years of trapping we watched the reaction of each bait bird in the bal-chatri; we sat and watched every set, waiting for the hawk to come in. Harry Meinel sometimes read comic books during the longer waits, which we thought very slapdash of him.

But in the winter of 1959, we learned that we could leave hawk traps unattended. It came about in this way. We were trapping and banding prairie chickens, and it began to look as though an invasion of goshawks, and other large raptors, were going to put us out of business. Those in authority announced, "This won't do."

For many years, raptors were more my bailiwick than Frederick's. We were both interested, but as I worked for the Wisconsin Conservation Department 60 percent of the time and Frederick full time, I had more time for such things as hawk trapping. In the current crisis it was both logical and practicable for me to take on the problem.

Oddly enough it was a rough-legged hawk that gave us the most trouble in the early winter. One single roughleg, considered a "mouse hawk," harassed the prairie chickens in traps at one station till they scalped themselves; and sometimes the "mouse hawk" managed to kill them outright through the netting.

(In the early 1930s Frederick and I had firmly believed that hawk shooting was a necessary part of game management, but now, over twenty-five years later, we couldn't imagine ourselves shooting a hawk for "vermin control"! It was inconceivable.)

A day or two later, after doctoring another casualty, it was no longer inconceivable. If a feral dog learns to kill sheep, it is time to kill that dog. One rough-legged hawk was upsetting an enormous investment in research—the research that was to lead to saving the prairie chickens in Wisconsin. That hawk must go. We got out our guns.

Our prairie chicken traps were scattered over about twenty acres. We hauled a blind through deep snow into the field and took turns sitting in it with a loaded gun. A thermos of coffee and some bread and sausage kept our spirits up during the long, cold waits. After hours of sitting, gun in hand, Os Mattson watched a flock of chickens settle in at the far end of the field. Within moments the roughleg was upon them. At suppertime Os said, "It was just as though she had been waiting for them to come in."

Frederick is an inventive type. He asked, "Where can we buy some domestic chickens?" What a solution! If that hawk was "waiting for the chickens to come in," we could shorten her wait and entice her within range of the blind!

We bought some bantam roosters and put them in cages in front of the blind. It was my turn to sit in the blind, and I was so confident that the hawk would go for the banties, that I took my 20-gauge Parker shotgun (which I normally use for woodcock hunting) into the blind. Until now we had been carrying rifles, but it seemed silly to take a rifle for a hawk that was to come within twenty yards of me.

The roughleg ignored the banties.

In the meantime Berger and Mueller had been experimenting with bal-chatris considerably larger than the original 6-inch diameter

79

traps designed for kestrels. I said, "Dan, we are going to have to use unwatched traps—permanents—to protect the prairie chickens."

"I don't like it; the hawks will hurt themselves."

"We've got to try it. Look at our mortality on trapped prairie chickens!"

Berger and I built about ten quonset-shaped bal-chatris with 12-inch by 18-inch bases. Each prairie chicken station had about twenty traps all vulnerable to predation. But now each station also had one or two bal-chatris with lure birds working day and night. The flocks of chickens came in to feed (and get trapped) in the morning and then again in the afternoon, presenting rather short vulnerable periods. The permanents worked.*

We checked our traps three times a day and one by one, and *unharmed*, we caught the hawks that were endangering our project. We named them alphabetically, like hurricanes: Adeline, Bertram, Carolyn. . . .

If we couldn't transport them out of the area right away, we taped their tails to keep them feather perfect, and held the birds in our barn loft, until we could release them about 100 miles away.

* F. Hamerstrom and D. Berger. "Protecting a trapping station from raptor predation," *Journal of Wildlife Management* 26(1962):203–6. In fifteen years up to 1965, we banded almost 2,000 prairie chickens and sharp-tailed grouse. Thirty-three were killed in traps by predators. Of these, thirteen were killed in the winter of 1959.

Whenever we had banded most of the prairie chickens at one station, we moved all the traps to another.

Frederick sent Berger to set up a new station back of Pete Schildt's buildings. "Pete is friendly, but ask permission before you go into his field. I hope there'll be some kind of a trail, so you can drive in." (A tractor track can save hours of shovelling snow to get traps into position.)

When Berger returned, Frederick asked, "No trouble getting permission, was there?"

Berger replied, "I didn't have a chance to ask. Pete came out of the barn and do you know what he said to me? 'If I'd known you were coming, I'd have spread some manure.'"

"Nice welcome," was my laconic reply. I thought that was to be the end of it.

Six hawks ready for release.

Carrion mail

Caught hawks rest comfortably in tubes. They stay put, seldom struggle, and when we pull them out of the tube, they come out feet first and feather perfect.

Owls are another matter. Owls become active in darkness, do *not* find tubes soothing, and may even injure their wing joints fighting constraint. Sometimes we slip an owl into a tube for a moment or two to weigh it, but if we can't release it right away, we stuff the owl into a nylon stocking. This gentle straitjacket prevents injuries.

So we all started carrying nylons and discovered that they have many uses. A stocking is a good strong rope for tying equipment together, it is a handy container for carrying lunch, bands and pliers, or live birds, and makes picturesque emergency bandages. Our group spent much of its time banding on the 50,000 acre Buena Vista Marsh in Central Wisconsin, and we needed to have a way of finding each other to help pull a stuck car out of the mud or snow, to borrow equipment, or to meet from time to time. It was thus that Stocking Mail came into being. Stockings, containing notes, were hung in conspicuous places along trails and roads. The messages within tended to be straight to the point: "You took my pliers and didn't give them back. Bring them to Howett Silo.", or, "S.O.S. Bring pig quick. S. Hanson BG. P!" We weren't trying to write in code. This last message simply meant "I need a pigeon South of Hanson Booming Ground to catch a peregrine, and I need it now."

Keeping up the stocking supply was a problem. Frederick startled various ladies by asking for their used nylons. Paul dated a nurse and was able to supply us with an ample stock of white nylons with very few runs.

Frederick preferred to carry tan or beige nylons and, for some incomprehensible reason, selected the very best quality — ultra-sheer — to put in his pockets. He likes to travel first class, and does not even like to put raptors into cheap stockings.

He also has an imposing presence — perhaps because he is incapable of doing anything wrong. Once, when we were on an out-of-state collecting trip, a game warden stopped to see what we were up to. The official car roared into the driveway; an officer with a small reddish mustache, and a tight uniform, knocked loudly on the door. He paused when Frederick opened it. I think he was going to shout, "What's going on here?" but what he actually said was, "Excuse me, Sir, could I please look at your permit?"

Frederick invited him in, introduced me, and reached for his wallet. A delicate, sheer nylon stocking fell out of his back pocket and floated to rest on a patch of sunlight on the floor. The officer gasped and took a quick look at me. Was it mine? Plainly not!

Frederick stooped slowly, picked up the stocking, folded it carefully, put it back in his pocket, and handed the warden his collecting permit.

After finding that the permit was in order, the officer prepared to take his leave. Just before he reached the door, his small mustache twitched slightly. He gave a hopeful little cough and looked at the sunlit spot on the floor where the stocking had fallen. Perhaps an explanation was coming?

Frederick opened the door for him.

The famous Pony Express lasted only eighteen months. Stocking Mail among the hawk trappers has endured twenty-three years. Our next type of communication evolved spontaneously: in the early 1950s, Memorial Day weekend was reserved for banding peregrines along the Mississippi River. Groups from Illinois and Wisconsin converged, and often wanted to get in touch with each other before arriving at the established meeting place.

(It is interesting to note that the vast effort expended on our Memorial Day weekends for the love of the birds of prey is the kind of thing the federal government *pays* people to do nowadays.)

Mueller, who was travelling in our car, said, "I wish we could intercept the Rockford Bunch. They'll be travelling this same road." It seemed an idle wish.

A few miles down the road, Frederick, who was driving, automatically slammed on the brakes when he spotted a fresh road-killed woodchuck. Many of us had hawks to feed, birds flown in falconry, birds for behavior studies, or injured raptors that had been brought to us to nurse back to health. Of course I wanted that woodchuck to feed to my redtail. But Berger took it from me. He fastened a note to it and put it back on the road. The note simply said, "Meet you at the Cozy Kitchen in La Crosse for lunch."

We had just finished giving the waitress our orders at the Cozy Kitchen when we heard loud voices and bursts of laughter. The Rockford Bunch, bearing the note, had arrived.

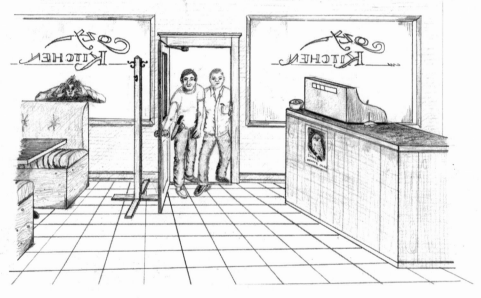

Not long afterwards the American Ornithologists' Union met in Denver. Berger knew that the Hamerstrom car was somewhere behind him. He had scooped up a dead jackrabbit, but there was no practical reason for sending us a note. Nevertheless he found a baby blue ribbon, fastened it around the dead rabbit's neck, and laid it in the driveway of the meeting place . . . where we couldn't miss it.

Berger spotted us as we were registering. "How'd you like my note?"

84

"Your note?"

"Yes, on that jackrabbit."

"We didn't see any rabbit."

Berger gasped, "Somebody else picked it up!"

Frederick asked, "What did the note say?"

"If I'd known you were coming, I'd have spread some manure."

Only drinking milk

Bird banders—almost without exception—are convinced that they are noble beings. Banding is not an end in itself; it helps solve the riddle of migration, makes possible the study of individual birds, and can help monitor environmental contaminants.

Bird watchers quite often feel noble too. Some of them are making first-rate contributions to ornithology (by conducting breeding bird censuses, for example). Others pursue rarities—sometimes to the point of harassment of endangered species—in their competitive eagerness to add just one more bird to their list. Aldo Leopold used to call running up a big list *bird golf,* and he viewed this undertaking with gentle tolerance. But that was long ago, before certain overly aggressive listers took bird golf too seriously.

The non-birdy public tends to lump banders and birders in an innocuous category: non-dangerous nuts. But if properly introduced to our occupations, they tend to buy bird books and even to think us quite wonderful.

The trouble with bird banding is that opportunities arise so swiftly that the few moments needed for a proper introduction to our work often are not there. But let me start with an outstandingly successful introduction. A snowy owl was perched on the roof of a split-level house in a Milwaukee suburb. It was close to the chimney, but up on the ridge pole, and looked ready to hunt. Two small children rode tricycles on the sidewalk making various motor sounds like *grrr grr grrr*.

We put two bal-chatris baited with pigeons on the front lawn. Then I rang the doorbell. A pleasant young woman in curlers came to the door. "Would you like to see something interesting?"

I wiped some of the mud off my boots and led her to her own picture window and pointed to the traps. "There's an owl on your roof and we're trying to catch it to band it."

"We've seen that owl. It chases cats. Are you going to let it go?"

"Yes," I added, "but those kids on tricycles are apt to scare it away."

My hostess was quick to act. "Why don't you invite the rest of your group in?"

She brought us steaming hot coffee and cookies, and disappeared to telephone. The children pushed their tricycles into the garage next door and disappeared too. The lady in curlers then telephoned all the neighbors to tell them to stay indoors and look out of their picture windows.

The snowy came off the roof in one swift glide. Our hostess gasped. It walked around the trap bobbing its head, and then it jumped up on the trap; tried to fly and was caught.

We rushed out of the house. Roughly thirty-five people rushed out of nearby houses to converge on our catch. We took the owl inside the bus to process it and the windows were so darkened by the mass of humanity that I had trouble reading the scales.

Not all encounters with the public turn out so well. In a nearby suburb a snowy hunted over an abandoned field staked out for real-estate development (such areas are often a boon to wildlife until growing cities spread and swallow them up).

We got our traps out in sight of the owl and retreated to the edge of the highway to await developments. The owl sat atop a TV antenna; a pigeon reposed peacefully inside one trap and a starling ran back and forth in another and—small skiffs of snow scudded across the weedy field.

The owl left its perch, banged into the starling trap, tumbled with the force of the impact, and started circling the trap, now and again jumping to its top in her attempts to reach the starling.

A man walked out of the back door of his house with a medium-sized dog. He was nearer to the owl than we were. Helmut muttered hopefully, "Just walking his dog."

But the man got closer and closer to the trap! The owl flew back to the antenna and we took off running hoping to head the man off.

We yelled, "Go away! It's ours! Leave it alone!"

The wind whisked our voices to nothingness until we were upon him and the starling whizzed off into space. The whole end of our bal-chatri had been cut out with wire cutters.

"*What*," demanded Helmut, "do you think you are doing?"

Not one of us was prepared for his answer.

"I saw the big bird trying to help the little bird out of the cage, so I thought I'd help too."

*   *   *

Our problems with the public are incredibly varied. Not a few people have tried to run over hawks caught on traps. We had one redtail shot on a trap, and a Cooper's clubbed to death by a trout fisherman. During the early years we were sharply criticized for letting the hawks go instead of killing them. One town chairman got so irate on this subject that he threatened to pull every black hair out of Dan Berger's beard.

Some people smash traps, but many more steal them. Even nice people seem to do this. I cannot understand it. If you find a purse, or a wallet, or binoculars, you know they belong to somebody, but traps with live lure birds in them seem to fall into a special category: Finders, keepers.

*Frank Renn
with Cooper's hawk.*

We tried using signs by our permanents but to do any good they had to be conspicuous — and they *attracted* the public. These pickers-up of traps never seem to look for a tag; they just take them home for the kids, I suppose. We learned to hide our sets well, but not always well enough.

One lovely day in May with a medium wind from the northwest, the trap I'd put near the ant hill was gone. I got on my hands and knees to see if I could tell in which direction the hawk had dragged it. I looked for tell-tale feathers or disturbed vegetation.

It would be a horrid thing to suspect that a caught hawk had dragged the trap to some secluded spot where it would go to its death if we couldn't find it. I worked over about half an acre and came home late for lunch, which I was supposed to have cooked. The crew was hungry. Right after a quick lunch we all looked for the lost hawk, hour after hour.

We argued. I said, "We ought to tie the traps to trees or bushes."

Frank insisted, "If we tie the trap, the hawk is apt to hurt herself, or she'll bust the nooses and get away."

Raymond declared, "I never lost a hawk yet. They can't drag 'em so far that I can't find 'em."

Two days later another trap had disappeared. Frank searched the area thoroughly. Then, instead of calling for help, he walked back along the sandy trail examining tire marks. He came in tired and late for lunch. "Somebody—none of our cars—drove in to the fallen oak set. That trap's got to have been pinched."

A few days later another trap, not far from a trail, was wantonly run over—smashed on purpose. These goings on were right on our farm, so when Dan Berger and I saw a car parked within sight of a fairly well concealed set for a goshawk, Berger floored the accelerator of his VW bus, swung in front of the parked car and stopped. The Muellers, in another VW bus, cut off the chap's retreat from behind so there was no way the driver could get away. Four of us descended on him.

"What are you doing?"

A scared-looking man, in worn work clothes looked at us with dazed watery eyes. He didn't seem to be able to find an answer.

"Where are you from?" I demanded.

"That farm up the road," he murmured almost inaudibly. "I'm the new hired man."

"Just what are you doing here?" Helmut roared.

The chap looked hopelessly to the front and to the back. There was no way he could escape. His jaw trembled and he slumped lower in the seat. "I'm . . . "

We all leaned forward to hear his explanation. This rattled him into silence.

At last he tried again. "I'm drinking milk. I have ulcers and I just stopped to drink milk."

And sure enough, he was clutching a quart carton of milk in both his trembling hands.

(We never did find the trap thief.)

On the art of Harris' hawk trapping

Trapping can never really be taught for it is an art. Good trappers have uncanny empathy not only for the species they are trying to catch but often for the individual they are after.

We have banded 652 Harris' hawks and retrapped thirteen, thus our catches total 665.*

Almost without exception our trapping has been by "road running." Roadrunners drive until they spot a hawk and then, usually from a moving vehicle, drop out a bal-chatri. For Harris' hawks we used starlings or feral pigeons.

When we first started trapping Harris' hawks we waited until we were sure the hawk was caught and then, full of joy, we rushed up to take the hawk off the trap for banding, etc.

We are wiser now.

When we've caught *one* Harris' hawk, we wait patiently for the friends to come in. Harris' hawks have rather a communal way of life. Two or three adults may tend one nest and partly grown young may be tolerated after the next family is on the way, so a nest with big young nearby may contain eggs!

This communal way of life is carried over into the hunting techniques of the Harris' hawks. We have watched one bird make a kill and watched the friends fly in to share the feast—and without fighting.

As I said, "We are wiser now." No longer do we rush in to take one hawk off the trap. We may pause for a cup of coffee, or tidy up the car. We wait for the friends to come—not to share the kill, but to try to—and so get caught on our traps. We have caught as many as four Harris' hawks on one 10 by 14 inch bal-chatri. We usually throw

From *Inland Bird Banding News* 49 (1977):4–8. Reprinted by permission. This article was coauthored by Frances and Frederick Hamerstrom.

* As of 1983 we have banded 1,091 Harris' hawks.

out two traps for each hawk and we have caught up to six Harris' hawks where we had set for only one. This is not as simple as it sounds, for several birds footing the trap and dragging it about mash the nooses down or pull them closed.

The scoop-up.

Footy feet dangerously near.

Everything under control.

Finally we invented the "scoop-up." Frederick is good at driving in roadside ditches. We must move rather fast or the caught hawks drag the trap away out of my reach at the critical moment: the moment when I crouch, lean out, and grab the bal-chatri with its adornment of beating wings, footy feet, and gyrating bodies. Just as I pull them all into the car, I toss out another trap with nooses standing up prettily — ready to catch the remaining friends. If at all possible we use the car as a screen so that the friends cannot see what we are up to, but this may just be thinking like a person. We know that stopping the car and getting out will spook the friends, but we suspect that the briefness of the interruption rather than what the hawks see is the critical factor.

We recently gave a talk to a group of biologists and described the scoop-up with considerable pride. A member of the audience asked, "Don't those birds ever get hurt? their wings or something?"

"No," I answered, "I'm the one who is apt to get hurt."

His reply was, "I'm very glad to learn that."

As Harris' hawks have footy feet, and the males especially are apt to take a notion to bite too, I thanked him somewhat tartly for his sweet compassion.

Night fright

Starlings are not all alike. Some are magnificent singers. We become fond of them. One of my starlings imitated the calls of yellow-headed blackbirds, redwings, western meadowlarks, and crowed like a domestic rooster. Another specialized in screaming like a red-tailed hawk on territory, imitated bluejays, and carolled the songs of unidentified thrushes.

Long trips at night are sometimes lonely. I drive through the dark countryside thinking my own thoughts, but whenever I pass through a lighted town, the starlings sing in the bait cage and I listen with delight.

Pigeons and starlings are surprisingly active at night. The sweet music of our bait birds, like the starry heavens and the silhouettes of nearby vegetation, gives us a sense of place in our camps. Pines and oaks designate familiar terrain; palms and cacti accentuate the lure of the exotic.

But sometimes we are not quite sure that we are camped in a safe place. It is astonishing how much misinformation well-meaning tourists can impart. In 1969 we headed for a trapping trip in the interior of Mexico: "Don't eat their food; don't drink the water; lock everything up—they'll rob you blind; I hope you are carrying a pistol?" Furthermore we were told that we would not be allowed to bring our bait birds into Mexico.

It was with trepidation that we approached the customs inspection. We had nineteen animals with us: three starlings, five sparrows, three mice, and eight pigeons. The American customs officials heaved a sigh and waved us on. It was different on the Mexican side. The officials, porters, and stragglers clustered around our vehicle, pointing at the pigeons and holding small children up to get a better look at them in the bait cage on top of the car.

A large official, with his belt pulled in tight over a bulging belly, kept a dignified distance and played the role of overseer, until I climbed up on top of the bus to scare the starlings out of the "house"

93

and into the cage where everybody could see them. Then the crowd parted, and he came closer.

In execrable Spanish I tried to explain the purpose of our expedition and what the birds were for. He listened intently, his plump cheeks raised in concentration.

Frederick had no idea of what the outcome of all this was going to be. I explained the birds, the mice, the traps, and the nooses. After I climbed down, the official turned to Frederick. "I see you have tricks." He spoke English!

Then he waved us on into Mexico.

We pulled as far off the road as we could to make camp, but over and over again, the local people discovered us and deluged us with presents. Near one seaside town, we were given so many presents that we had to leave sooner than we had intended, so as not to deplete our meager supplies by giving too many presents in return. None of these people looked dangerous to us.

Then we moved into the desert in Sonora. Hour after hour we followed a small, sometimes almost impassable road, and we saw no people, no houses, and no sign of recent tracks.

When the sun eclipsed down past a far purple mountain, we made camp. Frederick looked after the bait birds and rolled out the sleeping bag. I grilled our last piece of fresh meat over a small fire and warmed up some tortillas. We feasted, drank some red wine, and when night chill descended, we crawled into the sleeping bag.

Both of us are a trifle uneasy in cities, but lying under the stars in that camp we were at peace. It seemed as though we were the last people on earth.

Frederick stiffened in the sleeping bag. I woke and held my breath.

Siren! An ambulance was coming—nearer and nearer. "Whau-ah, Whau-ah, Whau-ah."

Where could we be? Frederick touched me reassuringly, but his touch was far from relaxed. The sound faded off into the distance.

The bait birds were restless. Pigeons cooed and starlings sang. The silhouette of a cactus loomed clear in the not quite dark night.

"What was it?"

Frederick gave me the quick squeeze that means: "I want to listen."

Then we heard it again. Far, far away, we could hear an ambulance—coming closer—"Whau-ah, Whau-ah, Whau-ah,"—and then gradually going away. Farther and farther away, until we could no longer hear it. "Do you think we ought to get up?" I whispered.

Frederick made no move.

Pigeons cooed, and a singer starling mimicked the strange grunting call of a yellow-headed blackbird that he had learned as a nestling raised near a marsh.

We made the connection simultaneously. The ghost ambulance was our singer ventriloquist: a starling that had been raised in a nest near a paved road on the way to a hospital!

"Whau-ah, Whau-ah, Whau-ah."

Winter visitors from the Far North

The owl's eyes are mere slits in the early afternoon sunlight; she has not moved. Two hours later she has not moved. The lake ice is a deeper blue and her eyes are becoming rounder. The shadows of the sedges around me lengthen and she rises — I can see half an owl. As she turns her head I see the golden yellow of her eyes. Suddenly she is up on her hunting perch and she sees the lure. My muscles tighten — ready for running. I wish we had more traps out and take two more out onto the ice.

Helmut has worked his way around the bay and we both watch. She's in like a flash and footing one of the traps. Oh, let her be caught! She jumps off, walking slowly around the trap as though pondering a way to get in (which she can't do). She coughs up a pellet of fur and bones beside the trap. She walks round and round the trap, bobbing her head. She's on the trap and foots it; but no, she isn't caught. Now she sees the next trap; in a long glide she reaches it and tumbles end over end. She tries for the inaccessible lure bird again and takes the trap with her this time. I fear she is just carrying the trap, not really caught.

Perhaps it is because snowy owls come from lands of drifting snow that they so often carry their food; if they cached it like other owls, they might have trouble finding it again in the shifting snow banks.

Our owl comes down now with the trap and walks around it again. She has lost interest in the first trap and we are glad we have several sets out. At sundown she looks up at one of the far sets and takes off again over the treacherous ice.

She hits the trap and foots it repeatedly, her white wings flashing against the dark ice. She's caught — dragging the trap! Helmut runs, but as I watch with the scope, he seems to go incredibly slowly, although I can see he has thrown caution to the winds and is running

This chapter is excerpted from *Audubon Magazine* 64(1962):12–15. Reprinted by permission.

directly toward her. He brings in the owl and we slip her in a nylon stocking where she will be safe and comfortable until we have time to examine her more thoroughly.

An owl caught on the Fond Du Lac dump. The pigeon—after a moment of panic—has resumed eating the corn in the trap.
Photo by Tom Meiklejohn

Helmut Mueller weighs a female snowy. She is gently restrained in a nylon stocking.
Photo by Orvell Peterson

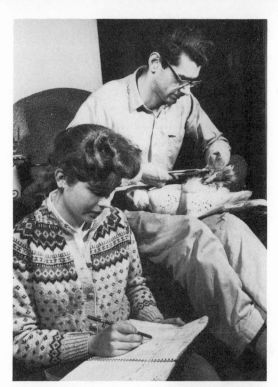

Elva Hamerstrom records while Dan Berger bands an owl.
Photo by Orvell Peterson

Spray-painting in a suburban basement. Left to right, Fran Hamerstrom, Dan Berger, Cynthia Schachter, Nancy Mueller.
Photo by Orvell Peterson

Fran Hamerstrom looks for fault bars on the wing of an adult female. Photo by Orvell Peterson

99

We marked the owls by dyeing or spray painting a portion of the bird's plumage, using colors such as pink, red, blue, green, yellow, black, or copper. This was primarily to get sight records, but it also turned out to be a great help in counting the owls, and in increasing our trapping efficiency. If we saw an owl with a red and green right wing, we didn't waste valuable hours trying to catch it, because we knew it was already banded.

Most of the color-marked owls that we had a chance to watch stayed in the same territories for weeks. It even appeared that these territories may not have been placid wintering grounds. Twice we saw female snowies towering high in the air with beating wings, presumably because one had invaded the other's territory.

Operation Snowy Owl was neither simple nor dull. I think of the hours spent watching owls on gas-storage tanks, coal piles, breakwaters, and housetops. Of hours in wind and snow, in city dumps among broken furniture and garbage, watching two traps and

A record catch. Left to right: Errol Schluter, Tanya Schluter,
Chuck Sindelar, Fran Hamerstrom, Paul Drake, Alan Jenkins.
Photo by Orvell Peterson

100

Night release. Photo by Orvell Peterson

a great white owl, motionless. I have memories of an Oshkosh owl, with dusk coming on, slowly bending its lovely head to feed on the chunk of prey between its feet, prey which it had been keeping warm for supper. We never caught it.

Twelve hours and twenty minutes was the longest we ever waited for an owl. We drove probably 200 miles on the ice and dropped our Volkswagens through it three times. We trapped on ice skates, from boats, on foot, and wading in hip boots through spring breakup. We trapped about forty-eight days, from December through March, and estimate that we drove a total of 8,634 miles.

Our trapping success was good. We believe that we had traps in sight of sixty-nine snowies, of which we caught fifty-eight. They were banded with United States Fish and Wildlife Service lock-on bands, as some large hawks and owls can remove ordinary bands.

Band returns may come in from the Far North. Perhaps an Eskimo, astonished to see an orange-winged owl, will catch it and find it banded. I have heard that the older Indians in the Hudson Bay area, who can no longer move around much, trap snowy owls, and that some of the Indians have as many as 150 as a winter food cache.

As I look back on our winters of banding snowies, I realize that our publicity campaign caused thousands of people to appreciate these spectacular birds.

Some species—equally worthy but modest brown birds—like the Ipswich sparrow, are hard to do anything for. Through no fault of their own, they receive only a moderately good press: they lack charisma. Just one snowy owl, seen once, can leave an indelible memory.

And then the snowies came

Driving on the treacherous ice of Wisconsin's lakes and marshes was a new experience for all of us. There were plenty of well-worn trails made by ice fishermen on the way to their shanties. The owls shunned these areas and perched on pressure ridges where the ice had buckled, or they hunkered, during the day, low on muskrat houses . . . sometimes not far from open water!

On Lake Mendota it was considered safe to follow the tracks of another car. The limnologists of the University of Wisconsin obeyed this rule and merrily followed a track clear into a hole where somebody else had dropped his car into the drink.

There was a large and to us exceptionally desirable snowy owl whose territory was near the Oshkosh causeway. Berger, who was trapping with Elva, asked the warden how near to it he could safely drive. The warden indicated a stretch of clear, even ice. Elva baited up the traps and Berger drove his vehicle (a white VW ambulance, equipped for trapping owls) nearer and nearer to the causeway female. She sat tight—head low and eyes mere slits—until that nice smooth ice gave way. All four wheels of the ambulance broke through at once. Suddenly afloat in the icy waters of Lake Butte des Morts, the ambulance rocked. Elva pulled off her boots and jumped out on the ice. Berger escaped. And the ambulance went to the botton.

By pure chance the bottom wasn't one of those thirty-foot horror holes. Berger's ambulance settled where the waters of this well-named lake managed to penetrate the motor with pernicious nicety. A wrecker hauled the vehicle out, but Berger never drove that ambulance again.

Nothing daunted, Berger showed Helmut and Nancy Mueller the causeway female the next weekend. This time they had Mueller's car, and Mueller inquired methodically about under-ice currents and danger spots. There was only one way to catch that bird: park on the busy causeway, let the traffic whiz by and walk some sets out to where she could see them. Cars roared past, brakes screeched from time to time. The causeway female remained placid and motionless.

It was not long before the police took notice. The owling party was ordered to move, get off the causeway and never park there again.

I missed that episode, but met the owlers at a nearby filling station the following Sunday and listened to their tale of woe. Then I asked, "Was it the state police?"

"Yes."

"OK, I'll call the sheriff."

Berger and Mueller tried to keep me from phoning. They insisted it was hopeless.

"Sheriff's department?" I put on my most official manner. "This is Dr. Frances Hamerstrom. There is an unusual invasion of snowy owls. We need to trap one near the causeway. . ."

My father was in police work—a criminologist—and I imitated his tone toward underlings—a tone that struck terror into all hearts, including mine. "We need an escort."

The answer was, "No."

"But it's a research project!"

To the amazement of everybody except me, a police car pulled into the drive and an officer in uniform walked into the filling station.

"Who's in charge?"

Nobody was in charge of Operation Snowy Owl, but I quickly said, "I am."

Ignoring normal introductions, I added, "Berger, you ride with the officer and show him where to park. We'll park right behind you."

Parked behind a squad car, we walked out on the ice and deployed our traps around the causeway female. Berger stayed in the squad car and enlightened the officer. He pointed out the need for banding to obtain data on movements, sex, age, and winter territories. He neglected to explain the color of his fingernails.

The officer kept looking at them, but it seemed he didn't like to ask personal questions. Berger, a lean outdoorsman, six foot three, with an intelligent face and broad shoulders, had brightly colored fingernails. Some were shocking pink, a few were yellow, and some were green.

*   *   *

104

Occasionally the officer peered in his rearview mirror to check on the inhabitants of our VW bus, but the windows were well steamed up. Cynthia, lowest in the peck order among gabboons present, had been instructed on how to start the camp stove and boil water.

"What's the water for?" she inquired politely.

Frank, one of the youngest members of our group, admired the ash on the end of his cigar and drawled, "We tell gabboons one thing at a time, so they don't have to take in too much at once. It keeps them from getting confused."

Our daughter Elva, a freshman at the university, once proudly stated, "I've never been a gabboon, but I want a gabboon of my own." That weekend she brought a friend home and so it happened that Cynthia was Elva's first gabboon—a slender, bright girl from downtown New York City. Cynthia could hardly suppress her excitement. When she asked, "What do gabboons do?" the other gabboons told her seemingly wild stories about rappelling down cliffs and going on expeditions to hunt bear. One took her aside and explained, "A gabboon is the lowest form of life. Scraps that nobody wants to eat are thrown to the gabboons. We have to clean the dirt out of the bait cages and all that sort of thing."

Cynthia didn't believe him. Elva had bestowed an honor on her: she was a gabboon.

Four of us sat in my VW bus, keeping an eye on the causeway owl. Two pots of water boiled on the camp stove. Finally Elva said, "Cynthia, would you please make us some hot chocolate?"

The gabboons shouted with laughter. "Elva said 'please' to a gabboon. What's the world coming to?"

Cynthia just thought they were being silly.

The crew in the bus was having a noisy good time. We especially enjoyed watching the state police drive past us as they cruised the causeway. Each time they went by, our officer gave his wrist a little flick to show that everything was under control and after a while they didn't even slow down when they went by. A red VW bus with a police escort had become a semi-permanent fixture on the causeway.

As time wore on, I got to feeling sorry for Berger sitting alone with our escort. "Elva, why don't you tell your gabboon to make another cup of hot chocolate. Then she can go and sit in the car with the police and Berger can come back here."

Cynthia obeyed. One of the gabboons opened the door for her and bowed deeply as she set forth for the police car with a cup of hot chocolate. "Stay forty minutes," I shouted after her.

In a moment or two Berger settled himself where he could watch the owl conveniently and said, "The officer is quite interested in birds. He has even bought himself a pair of binoculars which he keeps in the dash, and," Berger paused dramatically, "he hopes we won't give up, but stay right here until we get that owl."

When Cynthia came back she was fairly frothing with excitement. "I told him *everything* and . . . "

Not wishing to miss whatever was going on, I said, "I'll go next."

The officer started right in. "That's a very interesting young lady. I learned a lot."

Having no notion what he might have learned from Cynthia, I kept silent.

"I understand you run this show, Doctor?"

Having already informed the Sheriff's Department that I was in charge, I couldn't back out now. Besides I loved being called *doctor.*

"It seems that some of you are scientists and the others. . . ?"

"The Muellers are both scientists and my husband and I are both wildlife biologists. We work for the state—on prairie chickens. You should see them in spring! The cocks blow up huge orange air sacs," I spread my fingers on both sides of my neck to show where, "and they raise their neck tufts."

This man wasn't really listening to what I was saying. His eyes kept moving and he seemed to be watching my hands as though he were deaf.

Somewhat self-consciously, I put my hands on my lap and continued, "Sometimes as many as sixteen cocks come to a booming ground and they make a wonderful sound . . ."

He still wasn't listening so I stopped talking.

"What I wondered," he gave a small apologetic cough, "is what do the rest of you do? The gabboons?"

"Oh, one is a rodeo rider, some are in school, one works for Connie Mason—fixing chimneys."

"How about the man who was in the car with me first?"

"Berger. He's a good scientist and he runs a laundromat down in the Milwaukee slums."

"I didn't want to ask him . . . Now I see you do it too. Why do you paint your fingernails all those colors?"

It was on the tip of my tongue to say, "I don't," but my thumbnails glowed with shocking pink and more than traces of green, and a color called coppertone remained on the nails of most fingers.

"It's just accidental," I explained. "We don't do it on purpose."

106

"Accidental?" he gasped. Our officer was certainly listening to what I had to say now.

"Somebody has to hold the owls when we spray paint their feathers." I shrugged. "The paint gets on our hands. It washes off skin, but it doesn't come off fingernails."

"Well," he took a deep breath, "I hoped there was a reason."

This man in uniform wanted to ask me something more. And to my intense surprise, I noticed he was shy.

"I'd like to ask you a favor."

He moved his gloved hand lightly along the steering wheel.

"I'd like to be a gabboon."

The whole idea of a police officer gabboon was so mind-boggling that I crouched silent in my seat.

"Maybe I shouldn't have asked?"

"It's quite all right," I reassured him, "We'll give you a try — *on probation.*"

We never caught the causeway female.

Cynthia's slender body found the weekend unendurably cold and strenuous. Upon her return to college, she spent six weeks in the infirmary.

Operation Snowy Owl shifted to other cities so we never again had the pleasure of working with our favorite police escort.

Spring vacation. Our son Alan telephoned. "I have a ride all the way to Oshkosh. Can you pick me up there? About 2:00?"

"Wonderful. Let's meet in that restaurant next to Whitey's."

I got there first and glanced at a policeman sitting on a stool. To my delight it was my friend from the sheriff's department. Of course I settled next to him and told him about the owl in Milwaukee that was so tame that streetcars racketing right by didn't spook it from a utility pole, and about the owl that Helmut sneaked up on and hand grabbed near a coal pile. We talked . . .

When Alan arrived, I realized I had no idea what the officer's name was, but it didn't matter. I introduced them.

"This is my son, Alan."

They shook hands.

"And this is a gabboon."

Alan almost pushed me out of the restaurant and when the door swung shut, he sputtered, "How did you *dare* say that to his face?"

"Look here," I expostulated, "I made him *happy*. Up till now he was on probation to see if he could get to be a gabboon."

"He's a policeman!"

"Yes."

Alan wanted to drive. We sat in silence. Finally he asked, "Do I have this straight? You put a cop on probation to see if he would *qualify* as a gabboon?"

"Yes."

Accidents unlimited

Dr. Svein Myrberget was not fortunate. He arrived from Sweden to observe grouse, but Chuck Sindelar, a former gabboon, and I took him owling on Lake Butte des Morts—and we took him in Frederick's brand new VW bus.

We put our distinguished guest in the front seat, Sindelar drove, and I squatted unsociably in back repairing traps. But I *did* say, "Remember Chuck, the shoreline ice is thin."

We cruised open ice for some hours, and I only looked up near pressure ridges to try to spot an owl. Suddenly, at deep dusk I noticed the bus was hugging the shore of an island at the mouth of a channel. "Out!" I yelled, "The shore! Get away from the . . ."

With a nasty crunch and jerk, both right wheels broke through the ice. The bus listed sharply, hung at a slant and Sindelar jumped out the only door that would open—the driver's. Water poured into the bus. Svein followed Sindelar with dispatch and then I scrambled out by the same route carrying the unsheathed axe.

Sindelar and I laughed joyously—we were alive! Dr. Myrberget's ruddy complexion paled till his freckles stuck out like stalks; but he was game. Making strange small sounds like a mildly amused five-year old, he stated bravely, "If you can laugh, I can laugh too."

Sindelar and I took stock. We had a choice. We could either walk the safe, thick ice to a shore three miles away taking a chance on frostbite, or we could belly-crawl across a channel—the thin ice the car had dropped into, but perhaps we should trust it. Lights from a nearby house beckoned alluringly.

I flattened and snaked my way toward the light. "Didn't crack once," I shouted.

Svein followed, and then Sindelar, the heaviest, inched his way across.

The house had no telephone and a plump woman in a kimono directed us to a tavern some miles down the road. We trudged in darkness and beastly cold. When Sindelar finally spotted headlights of a car going our way, he pulled Dr. Myrberget into the shadow of a spruce saying, "Come with me." I pushed back the hood of my parka so a feminine face would show, and flagged the car down.

"Could you give us a ride to the tavern?"

"Pile in." The front seat was full and the back seat had room for just two more. Sindelar motioned Svein to get in. I gave Sindelar a little push; he squeezed into the back seat, and in a trice I was on his lap. Dr. Myrberget, who had never hitch-hiked before, spouted with laughter. "Very American . . . I am learning!"

Some four hours later, a rescue crew from the Oshkosh Fish Camp winched Frederick's new bus back up out of the water. The motor started without a sputter. Sindelar didn't know the nearest place to drive the car off the ice. He followed the fish camp wrecker as long as possible, but couldn't keep up. I knew the way, but I was in back again — this time feeding lure birds, drying out our notebook and wringing out wet clothing. A TV tower with a red light was my lighthouse in the dark. It seemed too close when I looked up from my duties. "Go right. You've got to turn south."

I found that Dr. Myrberget's briefcase was still up on the back seat, wedged between two traps, and dry. Then the headlights went out. Darkness! That's all I noticed. Both men jumped out of the car and I followed. There was a curious glow in front of the car. It seemed the headlights were not out after all. They were just both under water!

Utterly exhausted we abandoned that vehicle and walked across the ice till we could get a ride home. I was cheered by stars and a sparkling cold night. Frederick's car was not apt to drop through the rest of the way before tomorrow. More ice would be forming.

I was right. The next morning Sindelar and I had to chop ice to free the front end of the bus so the Fish Camp wrecker could pull us out again. Sindelar refused to drive. I can't blame him: both drop-throughs had been my fault, and Accidents Unlimited had been part of my life long enough so that I should have attended to guiding, rather than housekeeping.

That one day with Svein Myrberget cost me whatever standing I may have had as the most cautious member of Operation Snowy Owl.

I'll eat whatever's in it

The Cedar Grove Ornithological Station has played a major role in advancing research on birds of prey. One of the inventions of this ingenious group might never go down in history with proper credit if I failed to describe the evolution of the tube.

The Founders. Dan Berger and Helmut Mueller of the Cedar Grove Ornithological Station.

Methods for comfortably holding freshly caught hawks were undoubtedly developed by early falconers before Jesus Christ was born. One of the early methods, which I have seen performed, was to wrap the bird in a long scarf thus swaddling it like a Botticelli baby. The bird lies at rest, unable to escape and unable to injure itself. This is fine; but if you are catching twenty hawks an hour, it would take a whole crew just to do the wrapping.

The first step in the evolution of the tube was to acquire those long hollow cardboard stiffeners that carpets were once wrapped around for shipment. The stiffeners were sawed into appropriate lengths, one end was taped shut, and the prototype of the tube came into existence. It was in use for a very short period. Soon, two tin cans, one with one end removed and the other with both ends removed, were taped together. Air holes, punched around the edge of the closed end, provided ventilation.

It seemed at first that tin cans came in all the necessary sizes. Redtails fitted into 2-pound cans, female harriers slipped into fruit-juice cans, males into Marvalube cans, "Famous Chocolate Cookies" supplied merlin tubes. Campbell's soup cans were too big for kestrels and sharp-shinned hawks. But one company in the United States put out the perfect can for these. It manufactured—or perhaps I should say concocted—a vapid, sweetish pop. We drank all we could stomach and hoarded the cans. Then I hit upon the bright idea of giving a party so more people could help us empty the cans. A group of young people from Chicago didn't find this drink too bad; I watched with delight as they gulped it and then went into the kitchen to get more.

Chicago was already recycling cans. One of my well-brought-up young guests had stamped his can flat on the floor and was reaching for another!

"Don't, please don't!"

So we had twenty-three cans at the end of that party. And then, not to my surprise, the pop company went broke. Our supply of cans was sharply cut off.

Hawk trappers do not take defeat easily. My next approach made a lot of sense to me. I explored a large supermarket with one end in mind: I looked for 10-ounce cans. I walked the long rows examining cans of soups, vegetables, sauces, spices, fish, meat, gourmet products, juices, and baby foods. And I kept bumping into the same clerk who must have been taking inventory or something of the sort.

At last he asked, "Can I help you find something?"

"Yes, a 10-ounce can. What comes in a 10-ounce can?"

Somewhat rattled, he suggested I ask the manager. So I accosted an intelligent-looking ex-football player type.

113

"What comes in 10-ounce cans?"

"What did you want to get, Madam? What product?"

"Do they still make 10-ounce cans?"

He hunched his massive shoulders and spread his big hands palm upwards in despair. "Who's *they?*"

"Anybody. Can't you see? I want the *can*."

"The can?"

"Yes, I'll eat *anything* that comes out of a 10-ounce can."

The manager shook his head slowly, avoiding my insistent eyes. "Madam, I sure can't help you!"

We caught more kestrels than anything else, and it got so that just about the handsomest present you could give another bander was two empty 10-ounce cans. By late summer the shortage was acute. Frederick and I went on a banding expedition in Canada. It was not until we were on our way back, passing through a small town in northern Alberta, that we hit the jackpot.

"Stop!" I cried. "Please stop!"

Much mystified, Frederick brought the car to a halt just beyond the town dump where the glint of a likely-looking can had caught my eye. I ran back pell mell and found our first trophy can.

Frederick never looks as though he belongs on a dump. He just naturally has the aura of one who would tell somebody else to throw trash away, instead of grubbing in a dump to look for something for his own use. But he found almost as many *Pure Spring* root beer cans as I did — and a stout clean-looking cardboard box to put them in. We still had about 500 miles to look for *Pure Spring* cans before leaving Canada and by the time we got to the customs station we had two large boxes filled to the brim with fairly clean, empty root beer cans.

There is something uncomfortable about going through customs. Somehow, when one has the best of intentions, and no thought of smuggling, the officials have a knack for making one feel in the wrong. I brushed my hair, put on a clean shirt, and got braced for the inquisition.

After the usual polite questions, and very little poking about among our possessions, the man in the blue uniform seemed about to wave us on — but no.

"Just one minute. I'll take those boxes for you."

I realize now that he was being nice and wanted to dispose of our dirty rubbish, but at the moment it appeared that we were faced with a confrontation. I spoke firmly.

"We are *importing* those cans."

"Importing?" I can still see the look on that man's face.

The Royal Canadian Mounted Police

The United States police in general are attracted to anything unusual. Raptor trapping—until explained—draws the immediate attention of any normal police officer. Each encounter is different:

A squad car was parked next to ours.

"Is that car yours?"

"Yes."

"Why," demanded one officer, "are you picnicking on the town dump, instead of in the park?"

"Because there are rats here."

"What?"

"We are biologists. There are no rats in the park. They are right here on the dump. They attract hawks and owls, so this is where raptor trapping is good."

"What's your name?"

"Hamerstrom."

The officer looked at me as though I had just produced a magic password. "Why, that's right! Lady, we thought the Hamerstrom car was stolen." He added, "You could save us a lot of time if you'd let us know where you are going to be."

I rolled this unreasonable request over in my mind for a moment or two and settled for a simple apology. "I'm sorry."

Jack Oar tossed a baited bal-chatri out of his car to catch a kestrel. Just as he was about to turn around, a squad car roared up.

"Twenty-five dollar fine," the officer said. "I caught you littering."

"I wasn't littering," Jack protested innocently.

"I *saw* you littering! You just threw rubbish out of your car."

"It wasn't rubbish. It was a little cage with a mouse in it, and I'm going back to pick her up." Jack put his car in gear.

"You stay right here," the officer commanded sharply.

116

Jack peeked through the rearview mirror and saw that the kestrel was gone. The officer wrote down his license number, examined the plates, and came back to Jack's window.

"Now can I go back and pick up my mouse?"

"Your *mouse?*"

Jack got out of the car and started back along the edge of the highway.

"I'm coming with you." The officer spoke with authority. Together they walked the weedy shoulder of the highway until Jack said, "It ought to be about here."

"A *live* mouse?" The officer sounded as though he had a frog in his throat.

"Yes, I *told* you I wasn't littering."

Jack poked around in the vegetation, watched by a cop with arms akimbo.

"Must have gone farther than I thought," he mumbled.

Suddenly he stooped and picked up a bal-chatri within which a large white mouse, with long curved whiskers, raised its head inquiringly. Jack held her up for the officer to see. "Look, she's pregnant."

"Well," the officer sounded dubious, "I won't cite you. But you've put me to a lot of trouble."

"I'm sorry."

We were tired, dirty, travel-worn; we had grouse specimens in the car that must be mounted for the museum, and our laundry bag bulged with clothes we simply could not wear any more. As we drove into Fort Providence, a small town in the Northwest Territories, Frederick sighed. "I wonder if there's a laundromat?"

"Oh! Stop at the corner!" I shouted.

Then I jumped out of the car and asked a Mountie, standing nearby, "Is there a laundromat in town?"

Much to my mystification, he did not answer me. He walked around the front of the car and spoke directly to Frederick.

"Excuse me, sir, I'll be off duty in five minutes. I can help you then."

I got back in the car prepared to wait. Frederick asked, "What did you say to that man?"

"I just asked him if there was a laundromat in town."

Frederick looked impatiently at his watch. In exactly five minutes, the Mountie, who had paid no attention to us, turned and came back to Frederick's window. "If you'll just follow me, I can help you."

Frederick muttered, "This is ridiculous."

Preceded by our police escort, we drove three or four blocks to the edge of town and stopped in front of an attractive bungalow.

"I have a washing machine. There is no laundromat in Fort Providence."

He led us through an immaculate living room, then through a kitchen where one unwashed dish reposed in the sink. He coughed apologetically. "My wife is out of town; please forgive the mess."

The Mountie said, "I'll take that" and departed down a narrow flight of stairs into the basement with our laundry bag. I followed.

Frederick went back to the car to get our sleeping bag liner. Before long he started down the basement stairs. I think he was going to ask whether the liner was too big for the washing machine, but he stopped on the bottom step and said not a word.

I was standing perfectly still with my mouth open.

The Mountie, on his knees, moved with swift precision, reducing the contents of our laundry bag into two neat piles on the basement floor. Heavy outdoor clothes were tossed into one pile; my nighties and underwear were placed in another. He jumped to his feet and reached for the sleeping bag liner, which Frederick released in stunned silence.

After a moment of indecision, the Mountie laid the liner near, but not on, my lingerie, saying to Frederick, "You might wish to go upstairs and read."

But I had a better idea. "If it wouldn't be too much trouble, perhaps my husband could make bird skins on your kitchen table?"

"No trouble at all."

Frederick was soon happily at work saving specimens that we had hoped to skin earlier.

The Mountie stayed in the basement with me and regaled me with hair-raising stories of his work: how he did postmortems to find clues, etc. He was just helping me put the first load through the handwinding wringer, when the smell from the kitchen above became unmistakable: rotting grouse guts!

"I'm afraid we're making a frightful smell in your kitchen. Those grouse are over-ripe."

He carefully fed one of my pink rayon panties into the wringer and explained cheerfully, "Birds? I don't mind the smell of birds. It's bodies—long-drowned bodies—." He gulped. "I do not like *them*."

And so it went, and now Frederick and I were trapping in Canada—where Royal Canadian Mounted Police dressed in dashing uniforms rode well-groomed chargers. The travel ads had not quite prepared me for the real thing.

One day some sort of a large, almost black hawk flew toward the road and lit on a phone pole ahead of us. Frederick eased up on the accelerator, and I managed to get just one trap out as we passed the pole: a small heavy bal-chatri baited with a mouse. The hawk stayed put, and it seemed to us both that it must see the trap. A car that had just passed us made a swift turn, stopped at the trap and a chap got out and picked our trap up.

Frederick started the motor and launched full speed ahead to the rescue. I piled out of our car shouting, "That's *mine!*"

The man in uniform, not dressed as I expected a Mountie to be, tried to hand me the trap, but his fingers were caught in the nooses and he couldn't free himself. Thus it came to pass that we made the largest catch on record: A Royal Canadian Mounted Police Officer (estimated weight 186 pounds).

One yellow-bellied porcupine

An old sheepherder in Idaho taught me how to take care of camp meat. "You don't need ice. Hang it up at night when it's cold. Wrap it in your bed roll, and it will keep cold all day."

This information served us well on our expedition to the Northwest Territories. We could sleep free under the stars, and vegetables were cheap, but road-killed rabbits (our staple diet) were scarce in 1963. The scarcity of free meat ceased abruptly on September 25th: right on the busy highway near Emerson, Manitoba, lay a fresh-dead white domestic turkey weighing twenty-nine pounds. I scooped it up, plucked it by broadcasting feathers along the road toward Regina, and our meat supply was assured.

Frederick had gotten a little tired of having to stop for my repeated attempts to get supper off the road — especially as I had only been able to scoop up carrion, unfit for human consumption. So he was startled when I suddenly shouted "Porcupine. Darling, I *need* it!"

He continued on down the road. "It's bloated, and we have all that turkey."

"I need it for an Indian friend. She wants the fur for a headdress!"

Reluctantly he back-tracked and stopped at a particularly repulsive carcass.

Taking a quick whiff I explained, "I won't put it inside, I'll just put it on top."

The next city we went through some two days later was Edmonton. The porcupine was on the cartop carrier when we entered the city. It was gone when we left. Somewhere in traffic — in that sparkling clean city — we must have taken a corner too fast, and deposited that unlovely object with a plop.

Frederick tried to console me. In fact he spotted the next yellow-bellied porcupine before I did.

I had never skinned a porcupine so was pleased when we made a

121

deal with an old Indian named Margaret. Margaret was to take the fat off the skin and leave the hair on. My Winnebago friend wanted that extra-long hair of that yellow-bellied porcupine. (Wisconsin porcupines have shorter guard hair.)

When we returned the next morning, little wrinkled Margaret came half hobbling, half running up the hill toward us, her toothless smile full of delighted pride. The fat was still on the skin; the hair had all been scraped off! There are times to pay up even if people do the wrong thing.

The main object of this expedition was to study part of the range of the northern sharp-tailed grouse. We collected few specimens because we were two years late for the cyclic high, but I found a hatched nest eighty miles northeast of Fort Providence.

The Mackenzie Highway is the only road that leads to Yellowknife. Frederick sometimes explored east of the highway while I took the west. It was no time to get mixed up. To the south lay the Mackenzie River—so far to the south that one could not make it on foot before freeze-up; to the west lay the Alaska Highway some 500 miles away; and to the north lay the North Pole. No roads between me and the North Pole. . . .

Frederick, a natural woodsman, just took it for granted that because exploring separately was more efficient, that's the way we'd sometimes do it. I watched my compass. Compasses are not always to be trusted as bog ore throws the needle off, so I kept track of the wind and the sun—just to make sure. Sometimes, after chasing a flock of grouse, I'd sit down to try to think back: did I really start out to the west this morning? If not, how do I plan my way out?

Frederick would have been considerably startled if I had said, "Wouldn't it be more fun to work together?"

I never asked that question, and I always found my way back to the road and the car.

Opportunities to buy gasoline were few. One lonely shack—many miles from the nearest town—had a driveway, a gas pump, and a sign: No Gas. A week or so later, on our way back south, the sign was gone so we pulled up next to the pump to buy gas. Immediately two rough-looking men came out of the shack.

"Come in. Come in and have some coffee."

Frederick explained that we'd just stopped to buy gas.

"Gas? There's no gas here."

122

"But your sign is down."

"Waal," the slightly more voluble of the old sourdoughs explained, "there's never bin any gas here."

We waited for further explanation. He scratched his ear. 'We put the sign up when we're out working in the woods, and take it down when we're to home."

"But why?"

"That way we get company!" He paused. "Jim, here, made a pie today," he added hopefully.

Apple pie, made from sun-dried apples is excellent. "Just slice the apples and put them out in the sun and leave them till you can still bend the pieces. They'll turn brown, so you put in a little extra cinnamon to hide the color."

Once they got talking, it seemed as though conversation had gone to their heads. (It reminded me of a long winter, snowed-in, in Wisconsin, and the heady excitement of the first visitor who made it through to us in spring.)

That night we slept in an abandoned borrow pit and the northern lights flared and leapt like giant tadpoles in the sky. A wolverine demolished a bal-chatri that we had left set overnight to catch a great horned owl. But those sourdoughs had reminded me of Wisconsin, and now no amount of beauty or excitement erased the knowledge that we were headed homeward without a yellow-bellied porcupine.

When Frederick noticed a soulful, far-away look in my eyes, as our camp fire died down, he asked, "What are you thinking about?"

It was time to evaluate our work, or just to savor the night.

My answer was one word. "Porcupine."

He shot me one the very next morning.

October 25. Cluny, Alberta. We have seen no rabbits dead or alive in the thousands of miles travelled since we left Portage La Prairie. Had porcupine for lunch fried with onions and bacon with hot pepper added. Served with beans. Delicious.

Our records document a transect of the massive crash of the rabbit population in 1963. An equally massive crash occurred in our finances.

October 28. Have less than $10.00 left.

October 28 was not a particularly good day. Frederick wore his belt twisted all day long. Late in the afternoon he swatted at a fly on the windshield and made a 13¼-inch crack.
He missed the fly.

Pow-Wow

I put the porcupine skin in the crisper when we got home. It takes about a week to get everything running smoothly after an expedition: ten minutes to store the perishables; three hours to store the equipment for the next trip; six days to answer mail, fill out reports, take care of specimens, spend some time with the children, and work over the field notes.

So it was that I didn't get to take the porcupine skin to Lydia until a sunny day in November. She held the skin up to the light bending it gently, and watched the sunlight frisk and twinkle over the long hairs with intense concentration.

Suddenly Chief White Rabbit came into the kitchen. "You bring that?"

I nodded.

He picked up the hide tenderly and turned it skin side up. "Did you skin that?"

The skin had not a single gleam of fat. It was slightly pliable because I'd rubbed Ivory Soap into it.

"Yes, I skinned it."

Chief White Rabbit laid the skin on the kitchen table. "How much do you want for it?"

"Why, nothing," I answered, surprised. "Lydia said she wanted to make you a roach."

I got up to leave. I wanted to get out of there, having the uneasy feeling that the White Rabbits were about to try to offer me something in return. This was a present.

I wasn't prepared for his next question.

"Do you like dancing?"

"Yes!"

Chief White Rabbit got out a road map and made a mark on it and wrote "May 30" in the margin. "I am inviting *you* to a Pow-Wow."

I took note of his slight emphasis on the word *you*. It never occurred to me to say, "My husband shot that porcupine. Can't he come too?" I just waited for May 30.

Winnebago. My costume lacked authenticity, but not glamor. I wore a long hand-woven Bavarian skirt, an embroidered Mexican Indian blouse, antique jewelry from Finland, beaded moccasins from the Northwest Territories, and a shawl. The shawl was an oriental table cloth I had inherited from my Grandmother Flint.

I drove over 200 miles that night, and I went alone.

When I got to the Pow-Wow, some of the young braves standing near cars at the roadside stopped me. "This is a private Pow-Wow. It is not for white visitors."

Somewhat rattled, I explained, "I'm invited. I'm dancing with the Winnebagos." They let me through. I parked my car and stood on a hillside humming to the drumbeat and moving my feet in the heel-toe rhythm of the Indian dancers below.

Chief White Rabbit found me. "Want to dance now?"

"Give me ten more minutes."

"You came alone? No girl friend? No husband?"

"Alone."

He nodded approval.

Then I danced. There were only two white dancers, and heaven only knows how many Indians.

Small strings of women danced in the dust of a natural amphitheater. One string made room to let me in. Shyly I concentrated to learn the steps. Self-consciously I tried to hold my shawl as the other women did. The braves, in magnificent costume with bells on their ankles, swerved in and out with a quicker step. Their headdresses were mostly of eagle feathers, some wore big bustles, and some wore roaches.

Frost hit that night. It was so cold that some of the men wore pants and shirts in the early evening, but most were in full regalia—their dark skins shining with sweat even after the temperature dropped below freezing.

Inter-tribal Pow-Wow: when the drums got into my blood I forgot being careful. I danced.

One old woman kept pretending that she had lost the string she was supposed to be in and kept clowning and trying to shove in here and there. When she tried for my spot, I had my elbow ready, with a broad grin, to shoo her away. The braves danced and the women kept the beat — rather like a Greek Chorus. The best singer was a Winnebago and the best dancer was a Pawnee.

He dropped an eagle feather and — dancing — stooped as though to retie a moccasin lace, and scooped the feather up.

I learned later that technically he would have to pay for that dropped feather, but he improvised so well that they let him off.

Corn Dance! The women's feet moved swiftly. *They* took over the beat. One by one single women were pushed into the center — nearer the drums — to solo.

When I felt the sharp jab in my ribs, I whirled toward the center dancing alone — but really no longer alone. I was one with the drums, with the singing, with the Indians and with the heavens.

I danced the soles out of my moccasins that night. And before daylight, I scraped the frost off my windshield and drove home — slowly — the drums still beating in my blood.

I have had the honor of dancing at an inter-tribal Pow-Wow, and I was invited to come again. I savor that glorious night with respect and mystery. Sometimes when I come upon a sandy trail leading into the jackpines I breathe faster. I remember the feathers, the bells, the swirling bodies and — like a daydream remembered — the drums move into my being.

Fran and some of the Rockford Bunch. Left to right—with boots on—Ron Kern, Rodd Friday, Jim Weaver, Frank Renn, and Tom Oar.

EPILOGUE

BIRDING probably takes on almost as many forms as love. There are "listers" and there are banders. Listers can pick up numerous nice tips from banders, many of whom are specialists.

Some delight in the sheer beauty of birds away from civilization, and others practice considerable ingenuity in attracting birds to their feeders.

I've heard tell of some very strange feeders. Many years ago someone was telling me how he bred horned owls in a pigpen. I never learned the end of that fascinating story because a determined friend came up and said, "I want you to meet a man who has a bird feeder."

"Just about everybody has a bird feeder!"

"Come."

It was thus that I met an amazing man who tied meat to a board and fed hawks!

Meat, put out for birds of prey on our farm, has attracted just one red-shouldered hawk, *but* my success with millet, suet, and sunflower seed to attract small birds is just about as good as anybody else's, and I can discuss squirrel guards and price of seed with a certain amount of erudition.

There are solitary birders, and those who like to go out in groups. I like to be alone when I watch birds. I delight in groups— indoors. It is as impossible for me to watch birds with a group as to play the piano and cook at the same time.

Some bird lister groups bother me. The cloak of nobility seems to descend upon their shoulders because they are outdoors breathing the fresh air and "doing no harm." Some assume that attaining a big list makes one a conservationist—a quaint fancy.

I am most drawn to birders who are trying to *do* something, and are sufficiently disciplined to learn about our world of birds.

Banding is only one form of purposeful birding, and I some-

129

times feel that comfy cloak of nobility descending upon *my* shoulders because I tend to work hard at it. Just writing these stories has made it plain to me that I am lucky, rather than noble. Banding has drawn me to the excitement of the cities where snowy owls sit on utility poles and watch the taxis whiz by; it has caused me to meet and work with people I would never have encountered otherwise; it has given me the opportunity to watch hundreds of banded raptors take off and fly away against hundreds of different cloud formations; and it causes memories to dwell in my being—memories of a hawk owl swooping to a spruce top—memories of a Pow-Wow, and feathers and bells and swirling bodies—memories of the arctic, the jungle, and the desert—and of so many beautiful birds, banded so we can learn more about them.